Boost SAT Score
Through More Practice with
Six Realistic Practice Tests
for
SAT Chemistry Subject Test

Dr. N. Chen

List of Contents

Preface

How to use the practice test tests?

The Chemistry Subject Test assesses your understanding of the major concepts of chemistry and your ability to apply these principles to solve specific problems. According to College Board, preparation of SAT II chemistry subject include the following.

- One–year introductory college–preparatory course in chemistry
- One–year course in algebra
- Experience in the laboratory

Many studies have found that practice test is the most efficient way to study. The primary purpose of this book is to help you to prepare the Chemistry Subject Test, but may also be used to evaluate your understanding of chemistry topics after taking chemistry course. The practice tests are designed to simulate the real test, and questions cover the topics specified by College Board for chemistry subject test.

The practice tests can be used to identify weakness and study the topic more specifically. For this purpose, you are supposed to finish the course works in both chemistry (laboratory) and basic mathematics mentioned above. Before taking the real test, take one practice test each time, and find your weakness and study the topic specifically. You need have your textbook on hand, and study the topic as necessary.

Topics in SAT Chemistry Subject Test

The list below are topics specified by College Board and percentage of related questions in each section. They are not in order of the textbook coverage of these topics. You may use this list to identify a specific topic as your weakness and study specifically. This becomes more effective when test date is approaching.

Structure of matter (25%)

- **Atomic Structure**, including experimental evidence of atomic structure, quantum numbers and energy levels (orbitals), electron configurations, periodic trends

- **Molecular Structure**, including Lewis structures, three–dimensional molecular shapes, polarity

- **Bonding**, including ionic, covalent, and metallic bonds, relationships of bonding to properties and structures; intermolecular forces such as hydrogen bonding, dipole–dipole forces, dispersion (London) forces

States of matter (16%)

- **Gases**, including the kinetic molecular theory, gas law relationships, molar volumes, density, and stoichiometry

- **Liquids and Solids**, including intermolecular forces in liquids and solids, types of solids, phase changes, and phase diagrams

- **Solutions**, including molarity and percent by mass concentrations, solution preparation and stoichiometry, factors affecting solubility of solids, liquids, and gases, qualitative aspects of colligative properties

Reaction types (14%)

- **Acids and Bases**, including Brønsted–Lowry theory, strong and weak acids and bases, pH, titrations, indicators
- **Oxidation–Reduction**, including recognition of oxidation–reduction reactions, combustion, oxidation numbers, use of activity series
- **Precipitation**, including basic solubility rules

Stoichiometry (14%)

- **Mole Concept**, including molar mass, Avogadro's number, empirical and molecular formulas
- **Chemical Equations**, including the balancing of equations, stoichiometric calculations, percent yield, and limiting reactants

Equilibrium and reaction rates (5%)

- **Equilibrium Systems**, including factors affecting position of equilibrium (LeChâtelier's principle) in gaseous and aqueous systems, equilibrium constants, and equilibrium expressions
- **Rates of Reactions**, including factors affecting reaction rates, potential energy diagrams, activation energies

Thermochemistry (6%)

- Including conservation of energy, calorimetry and specific heats, enthalpy (heat) changes associated with phase changes and chemical reactions, heating and cooling curves, entropy

Descriptive chemistry (12%)

- Including common elements, nomenclature of ions and compounds, periodic trends in chemical and physical properties of the elements, reactivity of elements and prediction of products of chemical reactions, examples of simple organic compounds and compounds of environmental concern

Laboratory (8%)

- Including knowledge of laboratory equipment, measurements, procedures, observations, safety, calculations, data analysis, interpretation of graphical data, drawing conclusions from observations and data

Skills Tested

The College Board also breaks down questions by skill on this Chemistry Subject Test:

Skill	Percentage of Test
Application of knowledge	45%
Synthesis of knowledge	35%
Fundamental concepts and knowledge	20%

Types of Questions

There are total of 85 questions in Chemistry Subject Test. They are broken to three types of questions, and explained as follow.

Type 1: Classification Questions

Questions 1–25: For these types of questions, a list of choices that applies to several questions. *Each choice can be used more than once or not at all.* Here's an example:

> **Question 4–6 refer to the following ionic species**
>
> (A) X^+
> (B) X^{2+}
> (C) X^{3+}
> (D) XO_3^{2-}
> (E) XO_4^{2-}
>
> 5. A type of ion found in sodium acetate
>
> 6. A type of ion found in aluminum oxide
>
> 7. A type of ion found in potassium phosphate

Answers:

4. A
5. C
6. A

In this case, choice A was used twice, choice C was used once, while choices B, D, E were not used at all.

Type 2: Relationship Analysis Questions

Questions 101–115: Each question is comprised of two statements that are connected to each other by the word "BECAUSE". There will be a special section labeled "Chemistry" in the lower left–hand corner of your answer sheet where you can fill in your responses to these questions.

To answer this type of question, the first step is to decide whether each statement is true or false. If both or either of the statements are false, you can ignore the Correct Explanation (CE) circle. If both statements are true, you need to examine if the second statement is appropriate explanation of the first one. In another word, only when (1) both statements are true and (2) the second statement is appropriate explanation of the first one, you fill the CE circle, and it's not always the case.

Here's an example:

101. Lithium atom is larger than fluorine atom	BECAUSE	Fluorine atom has more protons in its nucleus
103. A 1.0 M solution of HCl has a low pH.	BECAUSE	HCl contains chlorine.

Your responses may look like this:

	I		II		CE*	Diff. Level
101	●	Ⓕ	●	Ⓕ	●	3
102	●	Ⓕ	Ⓣ	●	○	3
103	●	Ⓕ	●	Ⓕ	○	3
104	●	Ⓕ	●	Ⓕ	●	4
105	Ⓣ	●	Ⓣ	●	○	5

For question 101, both statements are true, and the second statement properly explain the first one. Therefore, the CE circle is filled.

For question 103, both statements are true, but the CE circle is not filled since the second statement is not an appropriate explanation of the first one. The first statement is true because HCl is strong acid, and 1.0 M HCl must have a low pH because the high concentration of H^+ ion. Apparently, the second statement is also true, but the fact HCl contain Cl has nothing with the low pH (high H^+ concentration).

Type 3: Five–Choice Completion Questions

Question 26–85: These are "normal" multiple choice questions. They're stand–alone questions that simply ask you to choose the correct answer out of five choices.

Here's an example:

---

12. The relation between the pressure and the volume of a gas at constant temperature is given by

(A) Boyle's law
(B) Charles's law
(C) the combined gas law
(D) the ideal gas law
(E) none of the above

The answer is (A).

In some cases, you'll get a list of three statements labeled with roman numerals and will be asked to decide which one(s) is(are) true. Here is the example:

---

34. Which reactions would form at least one solid precipitate as a product? Assume aqueous reactants.

    I. $AgNO_3 + NaCl \rightarrow NaNO_3 + AgCl$
   II. $Pb(NO_3)_2 + 2KI \rightarrow PbI_2 + 2KNO_3$
  III. $2NaOH + H_2SO_4 \rightarrow Na_2SO_4 + 2H_2O$

(A) I only
(B) II only
(C) III only
(D) I and II only
(E) II and III only

Before looking at the answer choices, you need to go through each of the reaction equations and decide which ones form precipitation:

- Reaction I is true because AgCl is insoluble.
- Reaction II is true because PbI_2 is insoluble.
- Reaction III is not true because Na_2SO_4 is soluble.

Therefore, the right answer is (D).

Other Important Notes

- A periodic table and a list of abbreviation of commonly used chemistry terms are attached to the test. Therefore, you need to recognize what information you can find in a periodic table, what information you have to memorize.

- Calculator use not permitted. Problem solving requires simple numerical calculations. If you find a solution requires complicated calculation, it's an indication that you don't do it right and need to rethink your solution. However, you may not assume that all solutions are simple and straightforward, a complicated solution may end up with a simple calculation.

- Measurements are expressed in the metric system for all questions. There will not be any circumstance that you need to deal with English system, so don't try to transfer between metric and English system. If you have to do this, it's another indication that you may not do it right.

- You need to answer 85 questions in 60 minutes, approximately 40 seconds for each question. This means that you must answer the easy questions first, and come back to the more difficult questions later. A common mistake students make is trying to answer every question correctly, and spend too much time on the tough ones and have to give up some easy ones due to time limitation.

- Remember, you don't have to be right in every question to get a perfect score. What you need to do is to try answer as many question correctly as possible. If you have to give up some questions, you want to give up the tough ones or the ones you are not very sure about. This means that you may skip some questions during the exam, but remember to make marks and go back to these questions if you finish all other questions before the time is called.

How to Score the Practice Tests

The real SAT Chemistry Subject tests are scored by computer. You need to work out your score for the practice test in this book by following the procedure below:

STEP 1: Using the answer key on the next page of each test to determine how many questions you got right and how many you got wrong on the test.

Remember: Questions that you do not answer don't count as either right or wrong answers. This means that there is risk to guess.

STEP 2: List the number of right answers.

(A) _____

STEP 3: List the number of wrong answers here. Now divide that number by 4.

(B) _____ ÷ 4 = (C) _____

STEP 4: Subtract the number of wrong answers divided by 4 from the number of correct answers. Round this score to the nearest whole number. This is your raw score.

A) _____ – (C) _____ = _____

STEP 5: To determine your real score, take the number from Step 4 above, and look it up in the left column of the Score Conversion Table attached on the next page of the answer keys.

After you work out your practice test score

If you get a perfect score on the first practice test, congratulation, you are ready to take the SAT II Chemistry Subject test. You may want to take another practice test to confirm it. You may work out all remaining practice tests to keep your mind sharp before the real test.

If you don't get the expected score, don't be disappointed. Remember the primary purpose of this book is to help you study and get a good score in real test. So, check all the wrong answers and go back to the original questions to identify your weakness. Next, you want to examine the explanations attached at the end of each test, and evaluate your understanding of the topics. This will help you to study the textbook more specifically.

Furthermore, the average difficulty level of the practice test is slightly higher than the real test, so the score of these tests are not a reflection of the real test. If you can follow the procedure and study after each test, you will find that you improve after each practice test.

List of Figures

List of Tables

SAT II Chemistry

Practice Test 1

Answer Sheet

Part A and C: Determine the correct answer. Blacken the oval of your choice completely with a No. 2 pencil.

1	Ⓐ Ⓑ Ⓒ Ⓓ Ⓔ	25	Ⓐ Ⓑ Ⓒ Ⓓ Ⓔ	49	Ⓐ Ⓑ Ⓒ Ⓓ Ⓔ
2	Ⓐ Ⓑ Ⓒ Ⓓ Ⓔ	26	Ⓐ Ⓑ Ⓒ Ⓓ Ⓔ	50	Ⓐ Ⓑ Ⓒ Ⓓ Ⓔ
3	Ⓐ Ⓑ Ⓒ Ⓓ Ⓔ	27	Ⓐ Ⓑ Ⓒ Ⓓ Ⓔ	51	Ⓐ Ⓑ Ⓒ Ⓓ Ⓔ
4	Ⓐ Ⓑ Ⓒ Ⓓ Ⓔ	28	Ⓐ Ⓑ Ⓒ Ⓓ Ⓔ	52	Ⓐ Ⓑ Ⓒ Ⓓ Ⓔ
5	Ⓐ Ⓑ Ⓒ Ⓓ Ⓔ	29	Ⓐ Ⓑ Ⓒ Ⓓ Ⓔ	53	Ⓐ Ⓑ Ⓒ Ⓓ Ⓔ
6	Ⓐ Ⓑ Ⓒ Ⓓ Ⓔ	30	Ⓐ Ⓑ Ⓒ Ⓓ Ⓔ	54	Ⓐ Ⓑ Ⓒ Ⓓ Ⓔ
7	Ⓐ Ⓑ Ⓒ Ⓓ Ⓔ	31	Ⓐ Ⓑ Ⓒ Ⓓ Ⓔ	55	Ⓐ Ⓑ Ⓒ Ⓓ Ⓔ
8	Ⓐ Ⓑ Ⓒ Ⓓ Ⓔ	32	Ⓐ Ⓑ Ⓒ Ⓓ Ⓔ	56	Ⓐ Ⓑ Ⓒ Ⓓ Ⓔ
9	Ⓐ Ⓑ Ⓒ Ⓓ Ⓔ	33	Ⓐ Ⓑ Ⓒ Ⓓ Ⓔ	57	Ⓐ Ⓑ Ⓒ Ⓓ Ⓔ
10	Ⓐ Ⓑ Ⓒ Ⓓ Ⓔ	34	Ⓐ Ⓑ Ⓒ Ⓓ Ⓔ	58	Ⓐ Ⓑ Ⓒ Ⓓ Ⓔ
11	Ⓐ Ⓑ Ⓒ Ⓓ Ⓔ	35	Ⓐ Ⓑ Ⓒ Ⓓ Ⓔ	59	Ⓐ Ⓑ Ⓒ Ⓓ Ⓔ
12	Ⓐ Ⓑ Ⓒ Ⓓ Ⓔ	36	Ⓐ Ⓑ Ⓒ Ⓓ Ⓔ	60	Ⓐ Ⓑ Ⓒ Ⓓ Ⓔ
13	Ⓐ Ⓑ Ⓒ Ⓓ Ⓔ	37	Ⓐ Ⓑ Ⓒ Ⓓ Ⓔ	61	Ⓐ Ⓑ Ⓒ Ⓓ Ⓔ
14	Ⓐ Ⓑ Ⓒ Ⓓ Ⓔ	38	Ⓐ Ⓑ Ⓒ Ⓓ Ⓔ	62	Ⓐ Ⓑ Ⓒ Ⓓ Ⓔ
15	Ⓐ Ⓑ Ⓒ Ⓓ Ⓔ	39	Ⓐ Ⓑ Ⓒ Ⓓ Ⓔ	63	Ⓐ Ⓑ Ⓒ Ⓓ Ⓔ
16	Ⓐ Ⓑ Ⓒ Ⓓ Ⓔ	40	Ⓐ Ⓑ Ⓒ Ⓓ Ⓔ	64	Ⓐ Ⓑ Ⓒ Ⓓ Ⓔ
17	Ⓐ Ⓑ Ⓒ Ⓓ Ⓔ	41	Ⓐ Ⓑ Ⓒ Ⓓ Ⓔ	65	Ⓐ Ⓑ Ⓒ Ⓓ Ⓔ
18	Ⓐ Ⓑ Ⓒ Ⓓ Ⓔ	42	Ⓐ Ⓑ Ⓒ Ⓓ Ⓔ	66	Ⓐ Ⓑ Ⓒ Ⓓ Ⓔ
19	Ⓐ Ⓑ Ⓒ Ⓓ Ⓔ	43	Ⓐ Ⓑ Ⓒ Ⓓ Ⓔ	67	Ⓐ Ⓑ Ⓒ Ⓓ Ⓔ
20	Ⓐ Ⓑ Ⓒ Ⓓ Ⓔ	44	Ⓐ Ⓑ Ⓒ Ⓓ Ⓔ	68	Ⓐ Ⓑ Ⓒ Ⓓ Ⓔ
21	Ⓐ Ⓑ Ⓒ Ⓓ Ⓔ	45	Ⓐ Ⓑ Ⓒ Ⓓ Ⓔ	69	Ⓐ Ⓑ Ⓒ Ⓓ Ⓔ
22	Ⓐ Ⓑ Ⓒ Ⓓ Ⓔ	46	Ⓐ Ⓑ Ⓒ Ⓓ Ⓔ	70	Ⓐ Ⓑ Ⓒ Ⓓ Ⓔ
23	Ⓐ Ⓑ Ⓒ Ⓓ Ⓔ	47	Ⓐ Ⓑ Ⓒ Ⓓ Ⓔ	71	Ⓐ Ⓑ Ⓒ Ⓓ Ⓔ
24	Ⓐ Ⓑ Ⓒ Ⓓ Ⓔ	48	Ⓐ Ⓑ Ⓒ Ⓓ Ⓔ	72	Ⓐ Ⓑ Ⓒ Ⓓ Ⓔ

Part B: On the actual Chemistry Test, the following type of question must be answered on a special section (labeled "Chemistry") at the lower left–hand corner of your answer sheet. These questions will be numbered beginning with 101 and must be answered according to the directions.

PART B				
	I		II	CE
101	T F		T F	◯
102	T F		T F	◯
103	T F		T F	◯
104	T F		T F	◯
105	T F		T F	◯
106	T F		T F	◯
107	T F		T F	◯
108	T F		T F	◯
109	T F		T F	◯
110	T F		T F	◯
111	T F		T F	◯
112	T F		T F	◯
113	T F		T F	◯
114	T F		T F	◯
115	T F		T F	◯
116	T F		T F	◯

Periodic Table of Elements

Material in this table may be useful in answering the questions in this examination

1 H 1.0079																		2 He 4.0026
3 Li 6.941	4 Be 9.012												5 B 10.811	6 C 12.011	7 N 14.007	8 O 16.00	9 F 19.00	10 Ne 20.179
11 Na 22.99	12 Mg 24.30												13 Al 26.98	14 Si 28.09	15 P 30.974	16 S 32.06	17 Cl 35.453	18 Ar 39.948
19 K 39.01	20 Ca 40.48	21 Sc 44.96	22 Ti 47.90	23 V 50.94	24 Cr 52.00	25 Mn 54.938	26 Fe 55.85	27 Co 58.93	28 Ni 58.69	29 Cu 63.55	30 Zn 65.39	31 Ga 69.72	32 Ge 72.59	33 As 74.92	34 Se 78.96	35 Br 79.90	36 Kr 83.80
37 Rb 85.47	38 Sr 87.62	39 Y 88.91	40 Zr 91.22	41 Nb 92.91	42 Mo 95.94	43 Tc (98)	44 Ru 101.1	45 Rh 102.91	46 Pd 106.42	47 Ag 107.87	48 Cd 112.41	49 In 114.82	50 Sn 118.71	51 Sb 121.75	52 Te 127.60	53 I 126.91	54 Xe 131.29
55 Cs 132.91	56 Ba 137.33	57 *La 138.91	72 Hf 178.49	73 Ta 180.95	74 W 183.85	75 Re 186.21	76 Os 190.2	77 Ir 192.2	78 Pt 195.08	79 Au 196.97	80 Hg 200.59	81 Tl 204.38	82 Pb 207.2	83 Bi 208.98	84 Po (209)	85 At (210)	86 Rn (222)
87 Fr (223)	88 Ra 226.02	89 Ac 227.03	104 Rf (261)	105 Db (262)	106 Sg (266)	107 Bh (264)	108 Hs (277)	109 Mt (268)	110 Ds (271)	111 Rg (272)	112 (277)						

*Lanthanide Series

58 Ce 140.12	59 Pr 140.91	60 Nd 144.24	61 Pm (145)	62 Sm 150.4	63 Eu 151.97	64 Gd 157.25	65 Tb 158.93	66 Dy 162.50	67 Ho 164.93	68 Er 167.26	69 Tm 168.93	70 Yb 173.04	71 Lu 174.97

Actinide Series

90 Th 232.04	91 Pa 231.04	92 U 238.03	93 Np 237.05	94 Pu (244)	95 Am (243)	96 Cm (247)	97 Bk (247)	98 Cf (251)	99 Es (252)	100 Fm (257)	101 Md (258)	102 No (259)	103 Lr (260)

Note: For all questions involving solutions, assume that the solvent is water unless otherwise stated.
Reminder: You may not use a calculator in this test!

Throughout the test the following symbols have the definitions specified unless otherwise noted.

H = enthalpy	atm = atmosphere(s)
M = molar	g = gram(s)
n = number of moles	J = joule(s)
P = pressure	kJ = kilojoule(s)
R = molar gas constant	L = liter(s)
S = entropy	mL = milliliter(s)
T = temperature	mm = millimeter(s)
V = volume	mol = mole(s)
	V = volt(s)

Chemistry Subject Practice Test 1

Part A

Directions for Classification Questions
Each set of lettered choices below refers to the numbered statements or questions immediately following it. Select the one lettered choice that best fits each statement or answers each question and then fill in the corresponding circle on the answer sheet. A choice may be used once, more than once, or not at all in each set.

Questions 1 – 4 refer to the following

   (A) Sublimation
   (B) The triple point
   (C) The freezing point
   (D) The point of equilibrium
   (E) The boiling point

1. The combination of specific temperature and pressure at which solid, liquid, and gas phases exist simultaneously

2. The transition directly from the solid to the gas phase without passing through liquid phase

3. The temperature at which the vapor pressure of a liquid is equal to the pressure of the surroundings

4. The temperature at which solid and liquid phases exist simultaneously

Questions 5 – 8 refer to the following atoms/ion

   (A) Na^+
   (B) Al
   (C) Cl
   (D) Ti
   (E) Hg

5. Has electrons in f orbitals

6. Has seven valence electrons

7. Has an electron configuration $1s^2 2s^2 2p^6 3s^2 3p^1$

8. Has the same electron configuration as the neon atom

Questions 9 – 12 refer to the following

   (A) Chemical pH indicator
   (B) Acid/base buffer
   (C) Calorimeter
   (D) Electrolytic cell
   (E) Supersaturated solution

9. Has a cathode and an anode.

10. Addition of water to this solution will not change $[H_3O^+]$.

11. Change color at certain point when acid is added to base continuously.

12. Which is not in equilibrium.

Questions 13 – 16 refer to the following groups in periodic table

   (A) Halogens
   (B) Transition metals
   (C) Alkali metals
   (D) Alkaline earth metals
   (E) Noble gases

13. Have valence electrons in d orbitals.

14. Have the highest first ionization energies.

15. Exist as diatomic molecules at room temperature.

16. Need to lose one electron to form a stable octet.

GO ON TO THE NEXT PAGE

Questions 17 – 21 refer to the following types of bonds

    (A) s–s bond
    (B) s–p bond
    (C) p–p bond
    (D) Hydrogen bond
    (E) Metallic bond

17. Describes the bond between H_2O molecules

18. Describes the bond in F_2

19. Explain the electricity conductivity of Cu

20. Describes the bond in H_2

21. Describes the bond in HF

Questions 22 – 25 refer to the following expressions of amount

    (A) 9.03×10^{23} molecules
    (B) 22.4 liters
    (C) 3.5 moles
    (D) 10.0 grams
    (E) 6.02×10^{23} atoms

22. 0.5 moles of O_2

23. 66.0 grams of CO_2

24. 98.0 grams of N_2

25. 5.0 moles of H_2

GO ON TO THE NEXT PAGE

PLEASE GO TO THE SPECIAL SECTION AT THE LOWER LEFT–HAND CORNER OF PAGE 2 OF YOUR ANSWER SHEET LABELED CHEMISTRY AND ANSWER QUESTIONS 101–115 ACCORDING TO THE FOLLOWING DIRECTIONS.

Part B

Directions for Relationship Analysis Questions
Each question below consists of two statements, I in the left–hand column and II in the right–hand column. For each question, determine whether statement I is true or false and whether statement II is true or false and fill in the corresponding T or F circles on your answer sheet. *Fill in circle CE only if statement II is a correct explanation of the true statement I.*

EXAMPLES:

I		II
EX1. The nucleus in an atom has a positive charge.	BECAUSE	Proton has positive charge, neutron has no charge.

SAMPLE ANSWERS

	I	II	CE
EX1	● Ⓕ	● Ⓕ	●

	I		II
101.	Potassium permanganate is a colored compound	BECAUSE	both potassium and manganese are metals.
102.	Water molecules have polar covalent bonds	BECAUSE	the reaction of hydrogen with oxygen to form water is an exothermic reaction.
103.	Isotopes of an element have nearly identical chemical properties	BECAUSE	they have identical electron configuration.
104.	Boyle's law states that the volume of an ideal gas is proportional to its pressure	BECAUSE	as pressure increases on a gas, its volume decreases.
105.	A liquid can boil at different temperatures	BECAUSE	the vapor pressure of a liquid varies with the surrounding air.
106.	Chlorine has an atomic mass of 35.45	BECAUSE	chlorine has two isotopes with mass numbers of 35 and 37.

GO ON TO THE NEXT PAGE

107. An element that has the electron configuration $1s^2 2s^2 2p^6 3s^2 3p^6 3d^3 4s^2$ is a transition element **BECAUSE** in atoms of transition elements, the $1s$, $2s$, $2p$, $3s$, and $3p$ orbitals are completely filled in the ground state.

108. Atomic radii increase down a group **BECAUSE** the higher the atomic number within a group, the smaller the atom.

109. The bonds found in a molecule of N_2 are nonpolar covalent **BECAUSE** there is an equal sharing of electrons between the nitrogen atoms.

110. The empirical formula of glucose ($C_6H_{12}O_6$) is CH_2O **BECAUSE** the empirical formula shows the simplest ratio rather than the actual number of atoms in a molecule.

111. A salt whose water solution has pH of 5 is basic **BECAUSE** a solution with a pH of 5 has higher concentration of H^+ than OH^-.

112. Increasing the concentration of reactants will cause a reaction to proceed faster **BECAUSE** more reactants lowers the activation energy of a reaction.

113. Cl^- is the conjugate base of HCl **BECAUSE** a conjugate base is formed once a Brønsted–Lowry acid accepts a proton.

114. $F_2 \rightarrow 2F^- + 2e^-$ is a correctly written half reaction **BECAUSE** half reaction must correctly demonstrates conservation of mass and charge.

115. Ethane is considered to be a saturated hydrocarbon **BECAUSE** ethene has a triple bond.

GO ON TO THE NEXT PAGE

Part C

Directions for Five–Choice Completion Questions
Each of the questions or incomplete statements below is followed by five suggested answers or completions. Select the one that is best in each case and then fill in the corresponding circle on the answer sheet.

26. One mole of each of the following substances is dissolved in 1.0 liter of water. Which solution will have the lowest freezing point?

(A) NaCl
(B) $MgCl_2$
(C) $NaNO_3$
(D) Na_3PO_4
(E) CH_3OH

27. Which of the following equations is/are properly balanced?

   I. $Cl_2 + 2NaI \rightarrow I_2 + 2NaCl$
   II. $CH_4 + 2O_2 \rightarrow CO_2 + H_2O$
   III. $2K + H_2O \rightarrow 2KOH + H_2$

(A) I only
(B) I and II only
(C) I and III only
(D) II and III only
(E) I, II and III

$$C_2H_5O + O_2 \rightarrow CO_2 + H_2O$$

28. Ethanol and oxygen react according to the above equation.

How many grams of water can be produced from the complete combustion of 1.0 moles of ethanol?

(A) 22.5
(B) 45.0
(C) 67.5
(D) 90.0
(E) 11.2

29. A compound was analyzed and found to be composed of 75% carbon and 25% hydrogen. What is the empirical formula of this compound?

(A) CH_4
(B) C_2H_2
(C) C_2H_4
(D) C_2H_6
(E) C_3H_8

30. Which compound below has a tetrahedral molecular geometry?

(A) C_2H_4
(B) CH_4
(C) CO_2
(D) H_2O
(E) C_2H_2

31. Filtration is a technique particularly suited to the separation of

(A) a gas and a liquid
(B) two liquid with different densities
(C) two liquid with different molar masses
(D) a liquid and a solid
(E) two solids with different melting points

$$2NO(g) + H_2(g) \rightleftharpoons N_2O(g) + H_2O(g) + 351 \text{ kJ}$$

32. If the total pressure on the system is decreased when the reaction represented above is at equilibrium, which of the following occurs?

(A) The rate of the reaction increases.
(B) The concentration of H_2O increases.
(C) The temperature of the system increases.
(D) The concentration of N_2O increases.
(E) The ratio $[NO] / [N_2O]$ increases.

GO ON TO THE NEXT PAGE

33. Which letter in the boxes below has a value of 7?

Isotope	p	n	e⁻	Mass #	Atomic #
^{12}C				D	
^{13}C	A				E
^{14}N		B			
^{16}O			C		

(A) A
(B) B
(C) C
(D) D
(E) E

34. Which of the following compounds dissolves in water to form a strong acidic solution?

(A) Na_2O
(B) KOH
(C) $CaCl_2$
(D) SO_3
(E) CO_2

35. $H_2O(l) \rightarrow H_2O(g)$

Correct statements concerning the process above occurring at 100°C include which of the following?

  I.   The vaporization process is endothermic.
  II. The randomness of the system increase during vaporization.
  III. The average potential energy of the vapor molecules is greater than that of the liquid molecules.

(A) I only
(B) III only
(C) I and III only
(D) II and III only
(E) I, II, and III

36. What is the empirical formula of a compound that contains 0.25 mole of Ca, 6 grams of C, and 1 mole of O?

(A) CaCO
(B) $CaCO_2$
(C) $CaCO_3$
(D) $CaCO_4$
(E) CaC_2O_4

37. Enough solid AgCl is added to pure water at 298 K. After stirring the mixture for 15 minutes, the concentration of Ag^+ ions in solution is found to be 1.3×10^{-5} M. The K_{sp} value for AgCl is

(A) 2.6×10^{-10}
(B) 1.3×10^{-10}
(C) 1.3×10^{-5}
(D) 1.69×10^{-5}
(E) 1.69×10^{-10}

$$N_2(g) + 3H_2(g) \rightarrow 2NH_3(g) + 92 \text{ kilojoules}$$

38. Ammonium is produced from nitrogen by the exothermic reaction represented above. After 4 moles of N_2 are completely converted to ammonium, which of the following occurs?

(A) 184 kilojoules of heat are absorbed.
(B) 368 kilojoules of heat are absorbed.
(C) 92 kilojoules of heat are released.
(D) 184 kilojoules of heat are released.
(E) 368 kilojoules of heat are released.

39. If the pH of a solution is changed from 1 to 3 with the addition of NaOH, what percentage of H^+ ions is neutralized?

(A) 1%
(B) 10%
(C) 90%
(D) 99%
(E) 100%

GO ON TO THE NEXT PAGE

40. Of the statements below, which holds true for the elements found in Na_2HPO_4?

    (A) The percent by mass of sodium is 16.6%
    (B) The percent by mass of hydrogen is 8.3%
    (C) The percent by mass of phosphorus is 48.5%
    (D) The percent by mass of oxygen is 45.1%
    (E) None of above

41. Carbon and oxygen react to form carbon dioxide according to the reaction:

    $C(s) + O_2(g) \rightarrow CO_2(g)$

    How much carbon dioxide can be formed from the reaction of 32 grams of carbon with 64 grams of oxygen gas?

    (A) 36 grams
    (B) 64 grams
    (C) 44 grams
    (D) 132 grams
    (E) 88 grams

    $$2NO(g) + Cl_2(g) \rightleftharpoons 2NOCl(g)$$

42. What is the correct mass action expression (equilibrium constant, K) for the equilibrium represented above?

    (A) $K = \dfrac{1}{[NO]^2[Cl_2]}$

    (B) $K = \dfrac{[NO][Cl_2]}{[NOCl]}$

    (C) $K = \dfrac{[NOCl]}{[NO][Cl_2]}$

    (D) $K = \dfrac{[NO]^2[Cl_2]}{[NOCl]^2}$

    (E) $K = \dfrac{[NOCl]^2}{[NO]^2[Cl_2]}$

43. Which of the following processes will DECREASE the rate of a chemical reaction?

    I.   Using more concentrated reactants
    II. Decreasing the temperature by 25 K
    III. Stirring the reactants

    (A) I only
    (B) II only
    (C) I and III only
    (D) II and III only
    (E) I, II, and III

44. Which solution is not expected to conduct electricity?

    (A) Sodium chloride, $NaCl(aq)$
    (B) Glucose, $C_6H_{12}O_6(aq)$
    (C) Potassium, $KBr(aq)$
    (D) Acetic acid, $CH_3COOH(aq)$
    (E) Sodium hydroxide, $NaOH(aq)$

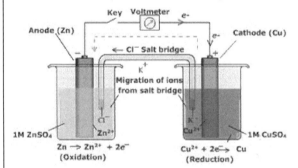

45. A voltaic cell is set up and a chemical reaction proceeds spontaneously. Which of the following does NOT occur in this reaction?

    (A) The electrons migrate through the wire connecting the two electrodes.
    (B) The cations in the salt bridge migrate to the anode half–cell.
    (C) The cathode gains mass.
    (D) The anode loses mass.
    (E) Reduction occurs at the cathode.

GO ON TO THE NEXT PAGE

46. What is the value for ΔH for the reaction:

$$A + B \rightarrow C + D$$

$A + B \rightarrow E$	$\Delta H = -390$ kilojoules
$E \rightarrow C + F$	$\Delta H = -280$ kilojoules
$F \rightarrow D$	$\Delta H = -275$ kilojoules

(A) −165 kilojoules
(B) +385 kilojoules
(C) −395 kilojoules
(D) −945 kilojoules
(E) +400 kilojoules

47. The oxidation state of the elements in the choices below is −1 EXCEPT for

(A) F in LiF
(B) Cl in HCl
(C) O in H_2O_2
(D) H in NaH
(E) H in Na_2HPO_4

48. Which substance is NOT paired with the correct type of bond found between the atoms of that substance?

(A) CH_4 — non polar covalent bond
(B) Li_2O — ionic bond
(C) NH_4^+ — polar covalent bond
(D) H_2O — polar covalent bond
(E) Cl_2 — non polar covalent bond

49. Which electron configuration shows that of an excited atom?

(A) $1s^2 2s^2 2p^6 3s^1$
(B) $1s^2 2s^2 2p^6 3s^2 3p^6 3d^1$
(C) $1s^2 2s^2 2p^3$
(D) $1s^2 2s^2 2p^6 3s^2 3p^6 3d^2 4s^2$
(E) $1s^2 2s^2 2p^6 3s^2 3p^1$

50. Given the chemical reaction

$$N_2(g) + 3H_2(g) \rightleftharpoons 2NH_3(g) + 92 \text{ kilojoules}$$

The forward reaction can best be described as a(n)

  I. Synthesis reaction
  II. Phase equilibrium
  III. Exothermic reaction

(A) I only
(B) III only
(C) I and III only
(D) II and III only
(E) I, II and III

51. Which of the following is NOT true regarding conjugates and conjugate pairs?

(A) HF and F^- are a conjugate pair
(B) NaCl and Cl^- are a conjugate pair
(C) CO_3^{2-} is the conjugate base of HCO_3^-
(D) NH_4^+ is the conjugate acid of NH_3
(E) Members of a conjugate pair differ by a transferrable proton.

52. Hydrogen bonding can be formed between a pair of molecules in solution listed below EXCEPT:

(A) H_2O and HF
(B) H_2O and H_2O
(C) NH_3 and NH_3
(D) NH_3 and H_2O
(E) CH_4 and H_2O

GO ON TO THE NEXT PAGE

22

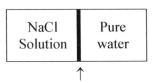

| NaCl Solution | Pure water |

↑
Porous Membrane

53. As shown in the diagram above, two compart– ments are separated by a porous membrane that is permeable to ionic salt. A solution is placed in one compartment while distilled water is placed in the other. Factors that influence the initial rate at which the NaCl diffuses into the compartment containing pure water include which of the following?

I. Concentration of NaCl
II. Area of the porous membrane
III. Temperature of the system

(A) I only
(B) II only
(C) I and III only
(D) II and III only
(E) I, II, and III

54. Weigh 118.9 grams of KBr, and dissolve in 500 mL of water in a 1 liter volumetric flask. Water is then added to make a total of one liter of solution. The final molarity of the solution is

(A) 4.0 M
(B) 2.0 M
(C) 1.0 M
(D) 0.5 M
(E) 0.595 M

55. Which of the following best describes the orbital overlap in a molecule of ethene?

I. s to s
II. s to p
III. p to p

(A) II only
(B) III only
(C) I and III only
(D) II and III only
(E) I, II and III

$$Cr_2O_7^{2-} + 6Cl^- + 14H^+ \rightarrow 2Cr^{3+} + 7H_2O(l) + 3Cl_2(g)$$

56. In the equation for the reaction represented above, which of the following indicates the reduction that takes place?

(A) $Cl^- + e^- \rightarrow Cl_2(g)$
(B) $Cl_2 + e^- \rightarrow Cl^-$
(C) $Cl^- \rightarrow Cl_2 + 2e^-$
(D) $2H^+ + O^{2-} \rightarrow H_2O(l) + 2e^-$
(E) $Cr_2O_7^{2-} + 14H^+ + 6e^- \rightarrow Cr^{3+} + 7H_2O(l)$

57. A titration is set up so that 100.0 mL of 1.0 M NaOH are titrated with 5.0 M HCl. If the initial reading of the acid's buret is 3.15 mL, what would the final buret reading be?

(A) 100.00 mL
(B) 40.00 mL
(C) 20.15 mL
(D) 23.15 mL
(E) 13.15 mL

GO ON TO THE NEXT PAGE

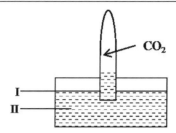

58. A test tube of carbon dioxide gas is inverted over water, as shown above. The gas become more soluble in the water if the

(A) Water is heated
(B) Gas is heated
(C) Lip of the tube is moved to level I
(D) Lip of the tube is moved to level II
(E) Amount of gas in the tube is reduced

59. The addition of diluted hydrochloric acid to an unknown metal produced a colorless gas. What is the likely identity of this gas?

(A) Cl_2
(B) O_2
(C) H_2
(D) CO_2
(E) NO_2

60. Which substance is NOT correctly paired with the bonding found between the molecules of that substance?

(A) NH_3 — hydrogen bonding
(B) F_2 — van der Waals (dispersion) forces
(C) HCl — dipoles
(D) CH_4 — dipoles
(E) NaCl(aq) — molecule–ion attraction

61. A 136 gram sample of a hydrated salt was heated at 150°C until all water was driven off. The remaining solid weighed 100 grams. From the data, the percent of water by weight in the original sample can be correctly calculated as

(A) $\frac{100}{136} \times 100$

(B) $\frac{36}{136} \times 100$

(C) $\frac{100}{236} \times 100$

(D) $\frac{100}{236} \times 100$

(E) $\frac{36}{236} \times 100$

62. When the equation :

$$\_\_C_3H_8 + \_\_O_2 \rightarrow \_\_CO_2 + \_\_H_2O$$

is balanced and the coefficients are reduced to the lowest whole numbers, the coefficient of O_2 is

(A) 4
(B) 5
(C) 6
(D) 8
(E) 10

$$^{112}Ag \rightarrow {}^{112}Cd + e^-$$

63. The decay of ^{112}Ag to ^{112}Cd represented above is a β decay with half–life of 3.13 hours. Given 100 grams of ^{112}Ag, how much ^{112}Ag is left after 6.26 hours?

(A) 100 grams
(B) 75 grams
(C) 50 grams
(D) 25 grams
(E) 12.5 grams

GO ON TO THE NEXT PAGE

64. Which of the following element has the greatest first ionization?

    (A) Ga
    (B) N
    (C) Ba
    (D) F
    (E) Ru

65. Based on the acid dissociation constants given below, it can be determined that which of the following is the strongest acid.

    (A) $HCN \leftrightarrow H^+ + CN^-$ $K_a = 4 \times 10^{-10}$
    (B) $HF \leftrightarrow H^+ + F^-$ $K_a = 6.3 \times 10^{-4}$
    (C) $H_2SO_3 \leftrightarrow H^+ + HSO_3^-$ $K_a = 1.4 \times 10^{-2}$
    (D) $H_3PO_4 \leftrightarrow H^+ + H_2PO_4^-$ $K_a = 7.1 \times 10^{-3}$
    (E) $H_2CO_3 \leftrightarrow H^+ + HCO_3^-$ $K_a = 4.4 \times 10^{-7}$

66. Which of the following would you NOT do in a laboratory setting?

    I.   Pour acids and bases over a sink
    II. Taste a solid to check if it is NaCl
    III. Heat a stoppered test tube

    (A) I only
    (B) I and II only
    (C) II and III only
    (D) I and III only
    (E) I, II, and III

67. An example of network covalence solid is

    (A) Table salt, NaCl
    (B) Dry ice, CO_2
    (C) Limestone, $CaCO_3$
    (D) Diamond, C
    (E) Glucose, $C_6H_{12}O_6$

68. Which of the following statements is NOT true regarding the kinetic molecular theory?

    (A) The molecules in a gas occupy no volume
    (B) There are no attractive or repulsive forces between the molecules
    (C) Collisions between gas molecules are perfectly elastic
    (D) Gas molecules travel in a continuous, random motion
    (E) The average kinetic energy of molecules is inversely proportional to temperature

69. Members of Group I elements have similar reactivity because they have

    (A) the same number of protons
    (B) the same number of electrons
    (C) similar outer shell configurations
    (D) valence electrons with the same quantum numbers
    (E) the same number

70. Which of the following reactions is NOT classified correctly?

    (A) $Fe + Cr^{3+} \rightarrow Fe^{3+} + Cr$ (redox)
    (B) $HBr + H_2O \rightarrow H_3O^+ + Br^-$ (hydrolysis)
    (C) $2C_2H_2 + 5O_2 \rightarrow 4CO_2 + 2H_2O$ (combustion)
    (D) $CH_4 + Cl_2 \rightarrow CH_3Cl + HCl$ (addition)
    (E) $CO_2 + H_2O \rightarrow H_2CO_3$ (synthesis)

STOP!

If you finish before time is called, you may check your work on this section only. Do not turn to any other section in the test.

Practice Test 1 Answers

PART A and C					
#	Answer	#	Answer	#	Answer
1	B	25	D	49	B
2	A	26	D	50	C
3	E	27	A	51	B
4	C	28	C	52	E
5	E	29	A	53	E
6	C	30	B	54	C
7	B	31	D	55	D
8	A	32	E	56	E
9	D	33	B	57	D
10	B	34	D	58	D
11	A	35	E	59	D
12	E	36	E	60	D
13	D	37	E	61	B
14	B	38	E	62	B
15	E	39	D	63	D
16	C	40	E	64	D
17	D	41	D	65	C
18	C	42	E	66	E
19	E	43	B	67	C
20	A	44	B	68	E
21	B	45	E	69	C
22	E	46	D	70	D
23	A	47	E	71	
24	C	48	A	72	

PART B	
#	Answer
101	True, True, No
102	True, True, No
103	True, True, Yes
104	False, True, No
105	True, True, Yes
106	True, True, Yes
107	True, True, No
108	True, False, No
109	True, True, Yes
110	True, True, Yes
111	False, True, No
112	True, False, No
113	True, False, No
114	False, True, No
115	True, False, No
116	

Calculation of the raw score

The number of correct answers: _____ = No. of correct

The number of wrong answers: _____ = No. of wrong

Raw score = No. of correct – No. of wrong x ¼ = _____

Score Conversion Table

Raw Score	Scaled Score	Raw Score	Scaled Score	Raw Score	Scaled Score
80	800	49	600	18	420
79	800	48	590	17	410
78	790	47	590	16	410
77	780	46	580	15	400
76	770	45	580	14	390
75	770	44	570	13	390
74	760	43	560	12	380
73	760	42	560	11	370
72	750	41	550	10	360
71	740	40	550	9	360
70	740	39	540	8	350
69	730	38	540	7	350
68	730	37	530	6	340
67	720	36	520	5	340
66	710	35	520	4	330
65	700	34	510	3	330
64	700	33	500	2	320
63	690	32	500	1	320
62	680	31	490	0	310
61	680	30	490	−1	310
60	670	29	480	−2	300
59	660	28	480	−3	300
58	660	27	470	−4	290
57	650	26	470	−5	280
56	640	25	460	−6	280
55	640	24	450	−7	270
54	630	23	450	−8	270
53	620	22	440	−9	260
52	620	21	440	−10	260
51	610	20	430		
50	600	19	420		

Explanations: Practice Test 1

1. **(B)** On a phase diagram, the **triple point** represents the pressure/temperature condition at which solid, liquid and gas of a substance can coexist (**Figure** 1.1).

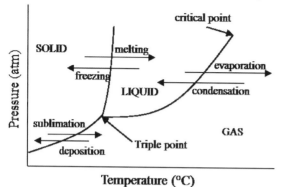

Figure 1.1 Phase diagram showing three states and a substance and processes associated with changes from one state to another. Lines are boundaries between phases on which two phases coexist.

2. **(A) Sublimation** is the transition of a substance directly from solid to gas phase without passing through an intermediate liquid phase (the reverse process is **deposition**). On a phase diagram, sublimation is represented by the arrow from solid to gas across the boundary between sold and gas (**Figure** 1.1).

3. **(E)** At **boiling point**, the vapor pressure of a liquid equals to the pressure of surrounding air. On a phase diagram, boiling point is any point on the line between liquid and gas (**Figure** 1.1). The diagram also shows that boiling point changes with when pressure, i.e. increase in pressure cause increase in boiling point.

4. **(C)** At **freezing point** (or **melting point**), the solid and liquid phases of a substance are at **phase equilibrium**. This is the temperature at which solid melts or liquid freezes without changing temperature under a constant pressure. Similar to Question 3, freezing point changes with pressure also (**Figure** 1.1).

5. **(E)** The electron configuration of Hg is [Xe] $4f^{14} 5d^{10} 6s^2$. Although Hg is in d section of the periodic table, it has filled f orbital at 4th shell ($n = 4$).

    Electron configurations of elements is generally considered a tough topic in SAT Chemistry Subject test. Student must understand the basic principles. A full understanding of the electron filling order (**Figure** 1.2 A) and periodic table blocks (**Figure** 1.2B) will help to handle most of the electron configuration questions.

    For this question, Hg is in 6th period. In this period, Ce (Z = 58) is the first element which has electron(s) filled to $4f$ orbital. Any element heavier than Ce has more than one electron filled to $4f$ orbital. Hg (Z = 80) is the only choice which is heavier than Ce, and has completely filled $4f$ orbital.

    Similarly, the first elements which has p electron is B (Z = 5), and elements heavier (Z > 5) have at least one p electron. The first element which has d electron is Sc (Z = 21) and elements heavier (Z > 21) have at least one d electron.

6. **(C)** All neutral atoms of **halogens** (F, Cl, Br, I, At) have 7 valence electrons.

7. **(B)** Al (Z = 13) has electron configuration of $1s^2 2s^2 2p^6 3s^2 3p^1$.

8. **(A)** Neutral Na atom (Z = 11, $1s^2 2s^2 2p^6 3s^1$) has one more electron than neon (Ne, Z= 10, $1s^2 2s^2 2p^6$). After it loses the 3s electron to become Na$^+$ ion, it has the same electron configuration as neon, and the electron configuration satisfies **octet rule**.

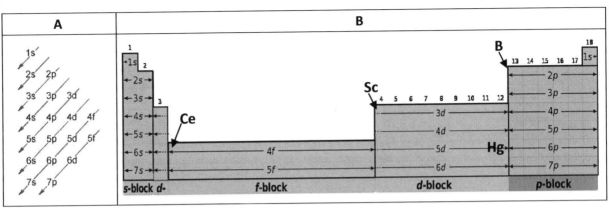

Figure 1.2 Order in which electrons are filled to orbitals (A) and electron configuration blocks of periodic table (B).

9. **(D)** Both voltaic (Galvanic) and **electrolytic cells** have **anode** (oxidation half–cell) and a **cathode** (reduction half–cell). However, the definition of anode and cathode are different in voltaic and electrolytic cells (**Table** 1.1).

<table>
<tr><td colspan="3" align="center">Table 1.1 Comparison between a Galvanic cell and an electrolytic cell</td></tr>
<tr><td></td><td>Electrochemical cell (Galvanic Cell)</td><td>Electrolytic cell</td></tr>
<tr><td>Diagram</td><td colspan="2">Reaction in a voltaic cell:
$Zn(s) + Cu^{2+}(aq) \rightarrow Zn^{2+}(aq) + Cu(s)$

Reaction in an electrolyte cell:
$Cu(s) + Zn^{2+}(aq) \rightarrow Cu^{2+}(aq) + Zn(s)$</td></tr>
<tr><td>Energy conversion</td><td>Converts chemical energy into electrical energy.</td><td>Converts electrical energy into chemical energy.</td></tr>
<tr><td>Reaction</td><td>The redox reaction is spontaneous and is responsible for the production of electrical energy.</td><td>The redox reaction is not spontaneous and electrical energy has to be supplied to initiate the reaction.</td></tr>
<tr><td>Cell setup</td><td>The two half–cells are set up in different containers, being connected through the salt bridge or porous partition.</td><td>Both the electrodes could be placed in the same container in the solution or molten electrolyte.</td></tr>
<tr><td>Electrode</td><td>The anode is negative and cathode is the positive electrode. The reaction at the anode is oxidation and that at the cathode is reduction.</td><td>The anode is positive and cathode is the negative electrode. The reaction at the anode is oxidation and that at the cathode is reduction.</td></tr>
<tr><td>Electron flow direction</td><td>The electrons are supplied by the species getting oxidized, moving from anode to the cathode in the external circuit.</td><td>The external battery supplies the electrons. They enter through the cathode and come out through the anode.</td></tr>
</table>

10. **(B) Buffer solution** can resist change in pH when small quantity of acid or base is added. In an acid/base buffer solution, there is a species which can act as both acid and base. When acid is added, it will absorb H_3O^+ ion to maintain the pH; when base is added, it will react with OH^- ion. An example is bicarbonate ion, HCO_3^-. In bicarbonate solution, there are following equilibria:

$H_2CO_3 + H_2O \rightleftharpoons HCO_3^- + H_3O^+$ (1) and $HCO_3^- + OH- \rightleftharpoons CO_3^{2-} + H_2O$ (2)

When acid is added into bicarbonate solution, equilibrium (1) will shift to the left to remove the added H_3O^+ ion. When base is added, the equilibrium (2) will shift to right to remove the added OH^- ion.

11. (A) **pH indicator** is a compound which show different color at different pH. An indicator usually is a weak acid/base (represented as HIn), and its acid form (HIn) and base form (In⁻) have different color. At low pH, it exists in its acid form, HIn; and at high pH, it exists as its base form, In⁻. When pH of a solution change from low to high or from high to low, as in an acid/base titration, the relative abundance of HIn and In⁻ will change, and the color of the solution will change when the solution change from HIn dominating to In⁻ dominating, and vice versa.

12. (E) A supersaturated solution has more solute is dissolved in solution. The system will shift to the direction to reduce the amount dissolved if the solution is disturbed, and solid will be formed out of solution to the point that the solution is just saturated. At this point, the solution reaches equilibrium between solid form and dissolved form of the solute.

13. (D) The elements of **transition metals** have electron(s) in unfilled d orbital. Electrons in the unfilled $3d$ orbital serve as valence electrons, rather than other than the outmost s orbital. This is because the outmost s orbital have lower energy than the unfilled d orbital. For example, Fe has electron configuration $1s^2 2s^2 2p^6 3s^2 3p^6 3d^6 4s^2$. The 6 electrons in $3d$ orbital are valence electrons rather than the two electrons in $4s$ orbital. As showed in **Figure** 1.2, transition metals are in d–block because the electron(s) in unfilled d orbital is(are) valence electrons.

14. (B) **First ionization energy** is one of the periodic properties of elements. Defined as the energy needed to remove one electron from a neutral atom, first ionization energy depends on how strong the nucleus attracts its outermost electron(s). In a periodic table, this attraction increases from left to right in a row (period). Therefore, the noble gas always has the highest first ionization energy in a row. The chart (**Figure** 1.3) below also shows some other periodic properties: electron affinity, atom radius, metallic/nonmetallic character.

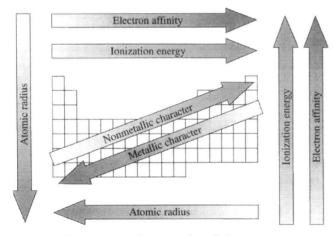

Figure 1.3 Some periodic properties of elements.

15. (E) All **halogen** elements exist as diatomic molecules at room temperature (i.e. F_2, Cl_2, Br_2, and I_2). As halogen atoms have 7 valence electrons, each atom share 1 electron with another atom to satisfy octet state. The formation of Cl_2 diatomic molecule is showed as follow:

:C̈l. + ·C̈l: ⟶ :C̈l:C̈l:

16. (C) On a periodic table, the elements in Group I (**alkali metals**) have one more electron than noble gases. This electron is in s orbital, and is the only valence electron. Alkali metals tend to lose this electron to

form a cation with one positive charge, M^+. The M^+ ion of alkali metals has the same electron configuration with a noble gas element. Similarly, the **alkali earth metals** in Group II will lose 2 valence electrons in s orbital to gain an electron configuration (M^{2+}) same as noble gases.

17. **(D)** Formation of **hydrogen bonds** between H_2O molecules makes it hard for water molecule to break the attraction from other water molecules to become vapor. In addition to elevate boiling point, hydrogen bonding also explain some unusual properties of water: high freezing point, universal solvent, lower density in solid state than liquid state, high heat of vaporization, high heat capacity high surface tension.

This set of question tests your understanding about **inter** *vs.* **intramolecular forces**. **Table 1.2** is a summary of inter– and intramolecular forces.

18. **(C)** In F_2 molecule, each of the two F atoms contributes one of $2p$ electrons to form covalent bond, and it's p–p bond with Lewis structure: $\overset{..}{\underset{..}{F}}\!:\!\overset{..}{\underset{..}{F}}\!:$ (see **Table 1.13**).

19. **(E)** In metallic bonds, electrons can move freely to conduct electricity (**Table 1.2**).

20. **(A)** In H_2 molecule, each of the two H atoms contributes its only $1s$ electron to form covalent bond, and the bond s–s bond with Lewis structure H:H (see **Table 1.13**).

21. **(B)** In HF molecule, H and F atoms form covalent bond as in Lewis structure $H\!:\!\overset{..}{\underset{..}{F}}\!:$. The H atom contributes its only $1s$ electron and F atom contributes one of its $2p$ electron to form a covalent bond. Therefore, the covalent bond between H and F is s–p bond (see **Table 1.13**).

Table 1.2 *Inter– and Intramolecular Forces*

Intermolecular forces

a. **London dispersion** (also called **van der Waals force**): the weakest intermolecular forces, which result from **instantaneous nonpermanent dipoles** created by random electron motion. It's the only intermolecular force between nonpolar molecules. London dispersion forces are present in all molecules and are directly proportional to **molecular size**. The bigger the molecule, the stronger the London dispersion force. For example, the diatomic molecules of halogens are all nonpolar, but their boiling point increases downward from F_2 to I_2, since the dispersion force increases with molecular mass.

b. **Dipole–dipole forces**: Dipole–dipole interaction occurs between two polar molecules such as H_2O, HCl, SO_2, CO etc. Polarity of a molecule is determined by the nature of intermolecular bond and geometry of the molecule. If all bonds in a molecule are nonpolar, the molecule must be nonpolar, for example, H_2, F_2, O_2, O_3, S_8. If the bonds in a molecule are polar, then you need to look the geometry of the molecule. Some molecules have polar bonds but the whole molecule is nonpolar due to shape of the molecule. For example, CO_2 (linear), BF_3 (trigonal planar), CH_4 (tetrahedral). Other molecules have polar bonds and the molecules are polar, such as HCl, H_2O (bent), NH_3 (trigonal pyramidal), etc.

c. **Hydrogen bonding**: Hydrogen bonding is the unusually strong dipole–dipole force that occurs when a highly electronegative atom (N, O, or F) is bonded to a hydrogen atom. Hydrogen bonding exists only in molecules with an N–H, O–H, or F–H bond, and explains some unusual properties of these molecules, such as their unusually high melting and boiling points.

Intramolecular forces:

a. **Ionic Bonds**: formed by the attraction of oppositely charged ions, for example, the bond between Na^+ and Cl^- ions.

b. **Covalent bond**: results from the sharing of electrons between two atoms with similar electronegativities. A single covalent bond represents the sharing of two valence electrons (usually from two different atoms).

Multiple covalent bonds are common for certain atoms depending upon their valence configuration. For example, a double covalent bond, which occurs in ethylene (C_2H_4), results from the sharing of two sets of valence electrons. Atomic nitrogen (N_2) is an example of a triple covalent bond.

The polarity of a covalent bond is defined by any difference in electronegativity the two atoms participating. Bond polarity describes the distribution of electron density around two bonded atoms. For two bonded atoms with similar electronegativities, the electron density of the bond is equally distributed between the two atoms. The **electron density of a covalent bond** is shifted towards the atom with the largest electronegativity. This results in a net negative charge within the bond favoring the more electronegative atom and a net positive charge for the least electronegative atom. This is a polar covalent bond.

A **coordinate covalent bond** (also called a dative bond) is formed when one atom donates both of the electrons to form a single covalent bond. These electrons originate from the donor atom as an unshared pair. Both the ammonium ion and hydronium ion contain one coordinate covalent bond each. A lone pair on the oxygen atom in water contributes two electrons to form a coordinate covalent bond with a hydrogen ion to form the hydronium ion. Similarly, a lone pair on nitrogen contributes 2 electrons to form the ammonium ion. All of the bonds in these ions are indistinguishable once formed, however.

Some elements form very large molecules by forming covalent bonds. When these molecules repeat the same structure over and over in the entire piece of material, the bonding of the substance is called **network covalent bond**. Diamond is an example of **network covalent bond**. In diamond, each carbon forms 4 covalent bonds to 4 other carbon atoms forming one large molecule the size of each crystal of diamond. Silicates, $[SiO_2]x$ also form these network covalent bonds. Silicates are found in sand, quartz, and many minerals.

c. **Metallic bond**: exists in pure metal substance such as copper, iron etc. In metal, valence electrons are **not** strongly bonded to particular atoms. This is called delocalization, and the delocalized electrons can move freely. This makes it possible for metal to conduct electricity.

22. (**E**) Since there are two O atoms in one O_2 molecule, 0.5 mol of O_2 molecules contain 1 mole of O atoms, which equals to 6.02×10^{23} O atoms.

*Questions are about conversion between mass, mole and number of particles, summarized in **Table 1.3**.*

23. (**A**) 66.0 gram CO_2 molecules = (66.0 gram CO_2)/(44.0 gram CO_2/mol CO_2) = 1.5 mol CO_2

   = (1.5 mol CO_2) x (6.02 x 10^{23} CO_2 molecules /mol CO_2) = 9.03 x 10^{23} CO_2 molecules.

24. (**C**) 98 grams N_2 = (98 gram N_2)/(28 gram N_2/mol N_2) = 3.5 mol N_2.

25. (**D**) 5.0 mol H_2 = (5.0 mol H_2) x (2 gram H_2/mol H_2) = 10 grams H_2.

Table 1.3 *Molecule and Mole*
Molecules
A molecule is the smallest unit of a compound that still displays the properties associated with that compound. A molecule is composed of two or more atoms that are held together by chemical bonds, such as covalent bonds and ionic bonds. The atoms of a molecule could be the same (as in O_2 and H_2 molecules), or different (as in CCl_4 and H_2O molecules). In the study of chemistry, the mass of a molecule is usually represented as its molecular mass, and the amount of a substance which is composed of the same molecules can be expressed as **mass** (gram) or **moles** (Avogadro's number).
Molecular Mass and Formula Mass
The molecular mass of a molecule is calculated by adding the atomic masses (in atomic mass units or amu) of all the atoms in the molecule. The formula weight of an ionic compound is calculated by adding its atomic weights according to its empirical formula. Example:
Molecular mass of H_2O = 2 x 1 + 16 = 18 (amu)   Formula mass of NaCl = 23 + 35.5 = 58.5 (amu)
Mole
Mole is a unit of measurement used in chemistry to express amounts of a chemical substance (e.g. atoms, molecules, ions, electrons). A mole of a substance is defined as the amount of that substance that contains

6.02 x 10²³ elementary entities, which equals to the number of atoms in 12 grams of pure carbon–12 (^{12}C). The number 6.02 x 10²³ is called Avogadro constant.

Calculations

Conversion of mass, mole and number of molecules of a substance is summarized as follow:

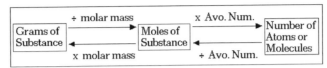

Examples:

1. How many H_2O molecules in 0.5 mole of H_2O?
    Solution: (0.5 mole H_2O) x (6.02 x 10²³ H_2O molecule/mole H_2O) = 3.01 x 10²³ H_2O molecules
2. When the word gram replaces mole in the problem above, you need to calculate the number of mole from the mass first.
    Solution: (0.5 gram H_2O) / (18 gram H_2O/mole) = 0.0278 mole H_2O
    (0.0278 mole H_2O) x (6.02 x 10²³ H_2O molecule/mole H_2O) = 1.67 x 10²² H_2O molecules

26. **(D)** The concentration of 1.0 mole of a substance in 1.0 liter of water is 1.0 M. In all salts given, 1.0 M $Na_3(PO_4)$ ($i = 4$) yield the highest concentration of ions (total 4.0 M of ions with 3.0 M of Na^+ and 1.0 M of PO_4^{3-}) in the solution, and will have the lowest freezing point (see **Table** 1.4).

    1.0 M of NaCl ($i = 2$), $NaNO_3$ ($i = 2$) yield solutions with 2.0 M all ions, and 1.0 M $MgCl_2$ ($i = 3$) yield solution with 3.0 M of all ions.

    CH_3OH ($i = 1$) is covalent compounds, 1.0 M CH_3OH solution cause least depression of freezing point.

Table 1.4 *Depression of Freezing Point*
When a non–volatile solute is dissolved in water, the **depression of freezing point** of the resulted solution depends on the concentration of the solution: $\Delta T_f = k_f \times m$ (1)   where ΔT_f is the observed change in the freezing point, K_f is the freezing–point depression constant (which is solvent–dependent), and m is the molality, or moles of solute per kilogram of solvent.   When the solute is a strong electrolyte, freezing point depression is also dependent on the concentrations of all ions in the solution: $\Delta T_f = i \times k_f \times m$ (2)   where i (Van't Hoff factor) is the maximum number of ions obtained from dissociation of the strong electrolyte. For example, the ideal Van't Hoff factor for NH_4Cl, $i = 2$, and for K_2SO_4, $i = 3$.

27. **(A)** Only equation I is properly balanced.

    II. is not properly balanced. Correctly balanced equation: $CH_4 + 3O_2 \rightarrow CO_2 + 2H_2O$.
    III. is not properly balanced. Correctly balanced equation: $2K + 2H_2O \rightarrow 2KOH + H_2$.

28. **(C)** The following is step–by–step calculation:

    Step 1: Balance the equation: $4C_2H_5O + 11O_2 \rightarrow 8CO_2 + 10H_2O$

    Step 2: Calculate number of moles of water produced. Based on the balanced equation, complete combustion of 1.0 mole of methanol will yield

    $$1.0 \text{ mol ethanol} \times \frac{10 \text{ mol } H_2O}{4 \text{ mol ethanol}} = 2.5 \text{ mol } H_2O.$$

Step 3: Calculate mass of H_2O: $2.5 \text{ mol } H_2O \times \dfrac{18.0 \text{ gram } H_2O}{1.0 \text{ mol } H_2O} = 45 \text{ gram } H_2O$.

29. **(A) Empirical formula** of a chemical compound is the simplest positive integer ratio of atoms present in a compound, what you need to do is to figure out the simplest **molar** ratio of the two types of atoms.

    Molar mass of carbon and hydrogen are 12 and 1 respectively. Assuming the **molecular mass** of the compound is M, the molar ratio of carbon to hydrogen is

    $$\frac{\text{Number of mol of C}}{\text{Number of mol of H}} = \frac{\frac{M \times 75\%}{12}}{\frac{M \times 25\%}{1}} = \frac{1}{4}$$

    Therefore, the empirical formula of the compound is CH_4.

30. **(B)** In CH_4 molecule, the valence electrons of C atom are hybridized (sp^3) to form 4 equivalent C–H bonds, and the CH_4 molecule has a tetrahedral geometry.

    (A) is incorrect because each C has triangular geometry. (C) CO_2 molecule has linear geometry. (D) H_2O molecule has bent geometry. (E) Both carbon atoms in C_2H_2 have linear geometry, and the molecule is linear.

31. **(D)** Filtration is suitable for separating solid (e.g. precipitate) and liquid. Solid is retained on filter, while liquid passes through the filter.

32. **(E)** This problem is about **Le Chatelier's Principle**, which states that if a chemical system at equilibrium experiences a change in concentration, temperature, volume, or partial pressure, the equilibrium shifts to counteract the imposed change and a new equilibrium is established (**Table** 1.5).

Table 1.5 *La Chatelier's Principle*	
Change	**Shift**
Increase concentration of a substance	Toward to direct to consume the substance
Decrease concentration of a substance	Toward to direction to produce the substance
Increase pressure of system	Toward to direction to produce fewer moles of gas
Decrease pressure of system	Toward to direction to produce more moles of gas
Increase temperature of system	Toward to absorbing heat to lower temperature
Decrease temperature of system	Toward to releasing heat to raise temperature
dd catalyst	No shift, but speed up reaction rate at both directions

For the reaction, $2NO(g) + H_2(g) \rightleftharpoons N_2O(g) + H_2O(g) + 351$ kJ, when pressure of the system is reduced, the equilibrium will shift to the direction at which the total volume of the system increases. The total number of moles in the left side and right side are 3 and 2 respectively; therefore, the equilibrium will shift to the left, and increase concentration of NO and decrease concentration of N_2O. This causes the increase in $[NO]/[N_2O]$ ratio.

33. **(B)** The **atomic number** (equals to number of protons) of ^{14}N is 7, hence the number of neutrons in ^{14}N is $14 - 7 = 7$.

Isotope	p	n	e	Mass #	Atomic #
^{12}C	6	6	6	(D) = 12	6
^{13}C	(A) = 6	7	6	13	(E) = 6
^{14}N	7	(B) = 7	7	14	7
^{16}O	8	8	(C) = 8	16	8

34. **(D)** is correct answer. Dissolution of SO_3 in water is expressed as $SO_3 + H_2O \rightarrow H_2SO_4$, and the product sulfuric acid (H_2SO_4) is a strong acid.

(A) $Na_2O + H_2O \rightarrow NaOH$ form a strong base
(B) $KOH + H_2O \rightarrow KOH$ form a strong base
(C) $CaCl_2$ is a neutral salt. When it dissolves in water, the solution is neutral.
(E) $CO_2 + H_2O \rightarrow H_2CO_3$ form a weak acid

35. **(E)** All three statements are correct.

I. Correct. Vaporization of liquid absorb heat, is endothermic process.
II. Correct. Randomness increases when liquid converts to gas.
III. Correct. Heat energy absorbed during vaporization is converted to the potential energy of water molecules in vapor.

36. **(E)** To find empirical formula, you need to convert the amount of each element in the compound to number of mole, and find the simplest whole number ratio following the steps below.

Step 1: convert all amount to number of mole. 6 grams C = (6 gram C)/(12 gram C/mol C) = 0.5 mol C.
Step 2: calculate the ratio: Ca : C : O = 0.25 : 0.5 : 1 = 1 : 2 : 4.
Step 3: write the empirical formula: CaC_2O_4.

37. **(E) Solubility equilibrium** exists when a chemical compound in the solid state is in chemical equilibrium with a solution of that compound. **Solubility product equilibrium constant** (K_{sp}) is the product of the equilibrium concentrations of the ions in a saturated solution of a salt. Each concentration is raised to the power of the respective coefficient of ion in the balanced equation. For a general dissolution reaction

$M_xN_y(s) \rightarrow xM^{y+}(aq) + yN^{x-}(aq)$ $K_{sp} = [M^{y+}]^x [N^{x-}]^y$

The K_{sp} of AgCl can be solved as follow:

a. Write the balanced dissolution equation: $AgCl(s) \leftrightarrow Ag^+(aq) + Cl^-(aq)$
b. At equilibrium, the concentration of $Ag^+(aq)$ and $Cl^-(aq)$ are the same, i.e.
   $[Ag^+(aq)] = [Cl^-(aq)] = 1.3 \times 10^{-5} M$.
c. $K_{sp} = [Ag^+(aq)][Cl^-(aq)] = (1.3 \times 10^{-5})^2 = 1.69 \times 10^{-10}$

38. **(E)** This is an exothermic reaction, meaning the reaction releases heat. Amount of heat released is calculated using stoichiometry as follow:

Heat released = (92 kJ/mol N_2) x 4 mol N_2 = 368 kJ.

39. **(D)** $[H^+]$ in solution of pH = 1 is $1.0 \times 10^{-1} M$ (0.1 M), and concentration of $[H^+]$ in solution of pH = 3 is $1.0 \times 10^{-3} M$ (0.001 M). The fraction of $[H^+]$ is neutralized by base is: (0.1 – 0.001)/0.1 = 0.99, or 99%.

40. **(E)** For this question, you don't have to calculate the precise percentage by mass of each element. You can make a estimation for each choice as follow:

First, calculate the molar mass of Na_2HPO_4: 2 x 23 + 1 + 31 + 16 x 4 = 112, and

(A) Total mass of Na in Na_2HPO_4 is 2 x 23 = 46. The percentage of Na in Na_2HPO_4 must be greater than 40%. Therefore, (A) is incorrect.
(B) Mass of H in Na_2HPO_4 is 1. The percentage of H in Na_2HPO_4 must be less than 1%. Therefore, (B) is incorrect.
(C) Mass of P in Na_2HPO_4 is 31. The percentage of P in Na_2HPO_4 must be less than 30%. Therefore, (C) is incorrect.
(D) Mass of O in Na_2HPO_4 is 4 x 16 = 64. The percentage of O in Na_2HPO_4 must be greater than 50%. Therefore, (D) is incorrect.
(E) Correct answer.

41. **(D)** This question is about the concept of limiting species. The following are common steps to solve such problems.

Table 1.6 *Limiting Reagent*
The **limiting reagent** is the reactant that is completely used up in a reaction and thus determines when the reaction stops. From stoichiometry, reactants are consumed at definite proportion. If the reactants are not supplied in the correct stoichiometric proportions (as seen in the **balanced chemical equation**), then one of the reactants will be entirely consumed while another will be left over in excess. The following are common steps in determining limiting reagent and amount of product formed, using the question in this problem as example. 1. Determine the balanced chemical equation for the chemical reaction. $C(s) + O_2(g) \rightarrow CO_2(g)$ ——— C and O_2 react at 1:1 ratio 2. Convert all given information into moles. 36 grams of C = (36 gram)/(12 gram/mol) = 3 mol C 64 grams of O_2 = (64 gram)/(32 gram/mol) = 2 mol O_2 3. Calculate the actual molar ratio from the given information. Molar ratio of C to O_2 given: 3/2 = 1.5 4. Compare the calculated ratio to the actual ratio. The actual molar ratio of C to O_2 is greater than the stoichiometric ratio which is 1, thus C is in excess, and **O_2 is limiting reagent**. 5. Use the amount of limiting reactant to calculate the amount of product produced. The amount of CO_2 produced is determined by amount of O_2, and is 2 moles. Since the molar mass of CO_2 is 44, the CO_2 formed is (44 gram CO_2/mol CO_2) x (2 mol CO_2) = 88 grams CO_2.

42. **(E)** is correct answer with appropriate exponents which are coefficients of reactants and products.

Table 1.7 *Chemical Equilibrium Constant*
For some chemical reactions, reactants are often not converted to product completely. In practice, many reactions reach a state of balance or **dynamic equilibrium** in which both reactants and products are present. A general chemical reaction at equilibrium can be expressed as aA + bB $\rightleftharpoons$ cC + dD where a, b, c, d are coefficients of reactants A and B, and products C and D. Under constant conditions (pressure and temperature etc.), the extent of the reaction is governed by the concentrations of both the reactants and products. The equilibrium constant, K_{eq}, is a measure of the extent to which reactants are converted to products in a chemical reaction at equilibrium. Equilibrium constants are found by multiplying the concentrations of the reaction's products raised to the power of their stoichiometric coefficients divided by the multiplication product of the concentrations of its reactants raised to the power of their stoichiometric coefficients. The equilibrium constant of the chemical equilibrium above is $$K_{eq} = \frac{[A]^a[B]^b}{[C]^c[D]^d}$$ where [A], [B], [C] and [D] are concentrations of A, B, C and D respectively. Under specified conditions, K_{eq} is always the same for a given reaction. It only changes when conditions (e.g. pressure and temperature) changes. This means that when you quote an equilibrium constant, you must specify the condition under which the equilibrium constant is measured.

43. **(B)** Using more concentrated reactants (I) and stirring the reactants (III) both increase the collision of reactants, and increase **rate of chemical reaction**. Only decrease in temperature (II) decreases reaction

rate. This question is about the factors that affect reaction rate. Some common factors and their impact on reaction rate are listed below (**Table** 1.8):

Table 1.8 *Factors Affecting Reaction Rate*	
Factor	**Impact**
Concentration of Reactants	For most reaction, higher concentration of reactants leads to more effective collisions per unit time, which leads to an increasing reaction rate.
Temperature	Higher temperature implies higher average kinetic energy of molecules and more collisions per unit time.
The Surface Area of solid or liquid reactants	Increase in surface area increases the collision of reactants. Example: using powdered reactants.
Presence of Catalysts	Catalyst lowers the activation energy of reaction, increase numbers of activated molecules which lead to formation of products.
Pressure	For gaseous phase reaction, increase in pressure increases the concentration of reactants, and increases rate of reaction.

44. **(B)** Aqueous solution of electrolyte conduct electricity. Glucose ($C_6H_{12}O_6$) is the only substance which is not an **electrolyte**. Acetic acid (D) is organic (weak) acid, and its aqueous solution is also electricity conductor. Other choices (A, C, E) are strong electrolytes.

Table 1.9 *Classification of Electrolytes*		
Electrolytes are substances which, when dissolved in water, break up into cations and anions. We say they ionize. Strong electrolytes ionize completely (100%), while weak electrolytes ionize only partially. That is, the principal species in solution for strong electrolytes are ions, while the principal specie in solution for weak electrolytes is the un–ionized compound itself.		
Examples of strong and weak electrolytes are given below:		
Strong Electrolytes	Strong acids	HCl, HBr, HI, HNO_3, $HClO_3$, $HClO_4$, and H_2SO_4
	Strong bases	**NaOH**, KOH, LiOH, $Ba(OH)_2$, and $Ca(OH)_2$
	Salts	**NaCl**, **KBr**, $MgCl_2$, and many many more
Weak Electrolytes	Weak acids	HF, **$HC_2H_3O_2$** (acetic acid), H_2CO_3 (carbonic acid), H_3PO_4 (phosphoric acid) etc.
	Weak bases	NH_3 (ammonia) etc.
Non–electrolyte	Molecule which does not ionize.	CH_3OH (methanol), CH_3CH_2OH (ethanol), glucose etc.

45. **(E)** is correct answer. In voltaic cell, oxidation reaction occurs at cathode, and reduction reaction occurs at anode. This is different from electrolysis cell (see **Table** 1.1).

46. **(D)** This question test your understanding of Hess's Law. **Table** 1.10 lists steps to solve this problem.

Table 1.10 *Hess's Law*
Hess's law states that the enthalpy change of an **overall** process is the sum of the enthalpy changes of its **individual** steps.
$$\Delta H_{overall} = \Delta H_1 + \Delta H_2 + \dots\dots + \Delta H_n$$
General procedure in calculating ΔH for an overall process
1. Identify the **target** equation, the step whose ΔH is unknown.

   o Note the amount of each reactant and product.

  2. Manipulate each equation with known ΔH values so that the target amount of each substance is on the correct side of the equation.

   o Change the sign of ΔH when you **reverse** an equation.

   o Multiply amount (mol) and ΔH by the same factor.

  3. Add the manipulated equations and their resulting ΔH values to get the target equation and its ΔH.

   o All substances except those in the target equation must cancel.

For question 46, the overall reaction equation can be obtained by adding three individual reactions without changing reaction direction and coefficients as:

$$
\begin{array}{lr}
A + B \rightarrow \cancel{E} & -390 \text{ kJ} \\
\cancel{E} \rightarrow C + \cancel{F} & -280 \text{ kJ} \\
+ \quad \cancel{F} \rightarrow D & -275 \text{ kJ} \\
\hline
A + B \rightarrow C + D & -945 \text{ kJ}
\end{array}
$$

47. **(E)** Use the oxidation rules in **Table** 1.11 to assign oxidation number (O.N.) for elements in this problem:

 (A) F in LiF -1 (Rule # 8)
 (B) Cl in HCl -1 (Rule # 8)
 (C) O in H_2O_2 -1 (Rule # 5)
 (D) H in NaH -1 (Rule # 4)
 (E) H in Na_2HPO_4 $+1$ (Rule # 10 and others)

Table 1.11 *Oxidation Rules*

Try to memorize the following oxidation rules as much as you can. Pay special attention to the **exceptions** as they are often the cause of mistakes.

1. In a formula, cation is written first followed by anion. *Example*: O.N. of H is -1 in NaH, but $+1$ in HCl.
2. The O.N. of a free element is always 0. *Example*: The atoms in S_8 and N_2, have O.N. of 0.
3. The O.N. of a monatomic ion equals the charge of the ion. *Example*, O.N. of Na^+ is $+1$; the O.N. of S^{2-} is -2.
4. The usual O.N. of hydrogen is $+1$. *Exception*: Hydrogen has O.N. of -1 in hydrides of active metals such as NaH and CaH_2.
5. The O.N. of oxygen in compounds is usually -2. *Exception*: (1) O.N. of O in OF_2 is $+2$ since F is more electronegative than O. (2) In peroxides such as H_2O_2 and BaO_2, O.N. of O atom is -1.
6. The O.N. a Group IA element (Li, Na, K, Rb) in a compound is $+1$.
7. The O.N. of a Group IIA element (Be, Mg, Ca, Sr, Ba) in a compound is $+2$.
8. The O.N. of a Group VIIA element in a compound is -1. *Exception*: (1) When that element is combined with one having a higher electronegativity. For example, the O.N. of Cl in chlorine monofluoride (ClF) is $+1$. However, in bromine monochloride (BrCl), O.N. of Cl is -1, and O.N. of Br is $+1$. (2) O.N. of Cl is $+1$ in HOCl, $+3$ in $HOCl_2$, and $+5$ in $HOCl_3$.
9. The sum of O.N. of all of the atoms in a neutral compound is 0.
10. The sum of O.N. in a polyatomic ion equals to charge of the ion. *Example*: The sum of O.N. for SO_4^{2-} is -2.

48. **(A)** Although the CH_4 molecules are nonpolar, the C–H bond in CH_4 is polar covalence bond. See Table 1.2 for details of different intramolecular forces (bonds).

49. **(B)** At ground state, $4s$ orbital has lower energy than $3d$ orbital, and should be filled before $3d$ orbital is filled (see **Figure** 1.2). An electron configuration of $1s^2 2s^2 2p^6 3s^2 3p^6 3d^1$ indicate that the $4s$ electron is excited to $3d$ orbital. All other choices are of ground state.

50. **(C)** The synthesis of ammonia from hydrogen and nitrogen gas is a synthetic reaction (I is correct), and the reaction release heat (exothermic reaction, III is correct). II is incorrect since all reactants and product in this reaction are gaseous. It's a **gaseous phase chemical equilibrium**, but not a **phase equilibrium**.

Phase equilibrium is an equilibrium between different phases, for example, equilibrium between liquid and vapor of water at boiling point.

51. **(B)** is the correct answer because NaCl and Cl^- are not a conjugate pair. As choice (E) correctly states, the **conjugate acid and base pair** differs by a transferrable proton (H^+). Therefore, (A), (C) and (D) are all correct.

52. **(E)** To form **hydrogen bond**, a hydrogen atom is attracted to an electronegative atom (proton acceptor) with lone electron pair(s) such as F, O and N; and the hydrogen must have enough positive charge to be attracted by the proton acceptor. Generally, hydrogen bonding can be formed between pairs of same or different molecules of HF, H_2O and NH_3. Therefore, hydrogen bond can be formed in pairs listed in choice (A)–(D). In answer (E), hydrogen in CH_4 is not positive enough to form hydrogen bond with H_2O.

53. **(E)** is correct answer. This question is about the factors affecting **rate of diffusion**. Concentration gradient, effective surface area of diffusion and temperature are among these factors (**Table** 1.12).

Table 1.12 *Factors Affecting Rate of Diffusion*	
Factors	**Impact on Rate of Diffusion**
Concentration gradient	The greater the concentration gradient, the faster the rate of diffusion.
Size of molecule	Smaller molecules have higher rates of diffusion than larger molecules.
Surface and permeability	Greater surface area and higher permeability increase rate of diffusion.
Temperature	Higher temperature (greater speed of molecules) results in higher rate of diffusion.
Pressure	Increase in pressure increases speed of the molecules and increases rate of diffusion.

54. **(C)** Follow the steps below to calculate the concentration of solution from mass of solute.

Step 1: Molar mass of KBr = 39 + 79.9 = 118.9 (g KBr/mol KBr).
Step 2: Number of mol in 118.9 g KBr = 118.9 g KBr/ 118.9 g/mol = 1.0 mol KBr.
Step 3: Concentration of final solution = 1.0 mol KBr/1.0 L = 1.0 mol/L (1.0 M).

55. **(D)** There are two types of bonds in ethane molecules. The C–H bond is *s–p* bond, and C=C double bond contains one *s–s* bond and one *p–p* bond (**Table** 1.13).

Table 1.13 *Valence Bond (VB) Theory*		
Valence Bond Theory assumes that all bonds are localized bonds formed between two atoms by the donation of an electron from each atom. According to VB theory, covalence bond is formed by sharing of orbitals between two atoms when the atoms are close enough to overlap their orbitals. A sigma (σ) bond is formed when bonding orbitals (*s* and/or *p* orbitals) of two atoms overlap end–to–end as found in single bond. A pi (π) bond is formed when bonding orbitals (*p* orbitals) overlap side–by–side as found in double and triple bond.		
σ *bond*: There are three types of σ bond. Depending on the types of valence electrons of the participating atoms, the σ bond can be formed between two atoms by overlapping two *s* orbitals (*s* to *s* overlapping), one *s* orbital and one *p* orbital (*s* to *p* overlapping), or two *p* orbitals (*p* to *p* overlapping).		
Type of σ bond	**Illustration**	**Example**
s–s bond	H + H → H H H· + ·H → H:H	H_2

s–p bond	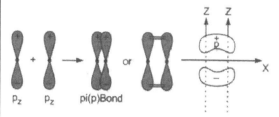	F_2
p–p bond		HF, HCl

π bond: π bond can be formed between atoms by overlapping two p orbitals side–by–side as illustrated below. π bond can be found in double and triple bonds. Remember, when more than one bonds formed between two atoms, the first bond is always σ bond, the second and third will be π bond. For example, a double bond can be considered to be consisted of one σ bond and one π bond; a triple bond can be considered to be consisted of one σ bond and two π bonds.

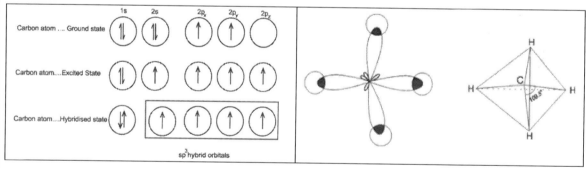

Orbital Hybridization

According to VB theory, carbon would only form two covalent bonds with hydrogen, making CH_2. However, this is not true and that in reality, formation of CH_4 is favored over CH_2. Orbital hybridization was introduced to solve this problem. Carbon is a perfect example showing the need for hybrid orbitals.

sp^3 hybridization: Carbon atom in the ground state has only two unpaired $2p$ electrons but in the excited state one $2s$ electron is promoted to vacant $2p_z$ orbital. Thus there are four unpaired electrons, one each in $2s$, $2p_x$, $2p_y$ and $2p_z$ orbitals. These four orbitals hybridize and form four equivalent sp^3 hybrid orbitals directed towards the four corners of a regular tetrahedron with bond angle 109.5°.

Each of the sp^3 hybrid orbital overlaps with $1s$ orbital of hydrogen atom and forms four C–H bonds in CH_4 molecule. Each C–H bond has s–sp^3 overlapping and methane molecule has tetrahedral structure.

sp^2 hybridization: In ethene (C_2H_4), each carbon atom is sp^2–hybridized. In this way six sp^2–orbitals are generated (three for each carbon atom). One sp^2–orbital of each carbon atom by overlapping forms a sigma bond between carbon atoms. Remaining two sp^2–orbital of each atom overlap with $1s$–orbital of hydrogen atom to produce four σ (s–p) bonds. The p_z un–hybrid orbital of each carbon atom by the parallel overlapping form a π (p–p) bond between two carbon atoms. Geometry of ethene molecule is trigonal in which bond angles are 120°.

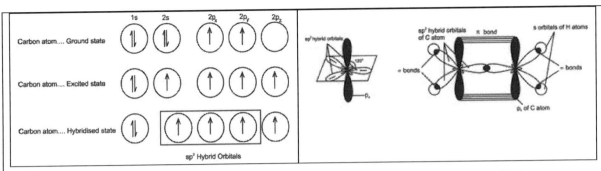

sp hybridization: In ethyne (C_2H_2), each carbon atom is *sp*–hybridized. In this way, four *sp*–orbital are generated. One *sp*–orbital of each carbon atom by overlapping forms a sigma bond between carbon atoms. Remaining one *sp*–orbital of each carbon atom overlap with $1s$–orbital of hydrogen atom to produce two sigma bonds. p_y–orbital of each carbon and p_z–orbital of each carbon by parallel overlapping form two pi–bonds between two carbon atoms. Geometry of ethyne molecule is linear in which bond angles are 180°.

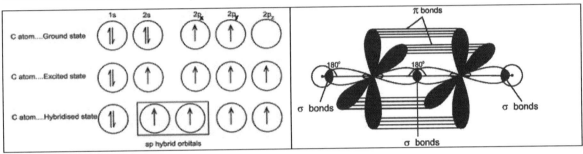

Other examples:

H_2O: oxygen atom in H_2O is sp^3–hybridized. Unlike carbon atom in CH_4, oxygen atom in H_2O has two more electron in the $2p$ orbital. Therefore, two of the four sp^3 orbitals are filled with two electron pairs, with the other two sp^3 orbitals form sigma bonds with hydrogen atoms. The angle between two O–H bond is 104.5°.

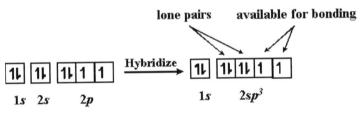

NH_3: similar to O atom in H_2O, N atom in NH_3 is sp^3–hybridized. Because N has one less electron than O, there is only one sp^3 hybridized orbital is filled with a pair of electron, while three sp^3 orbitals are filled with only one electron. These three half filled sp^3 orbitals can form three σ (s–p) bonds with H atoms.

BH_3: boron has 3 valence electrons ($2s^2 2p^1$), B atom in BH_3 is sp^2–hybridized, and form three σ (s–p) bond with H atoms.

56. **(E)** is correct answer. (A) is incorrect reduction half reaction. (B) is also a reduction half reaction but does not take place in the given reaction. (C) is correct oxidation half reaction taking place in given reaction (**Table** 1.14). (D) does not take place in the given reaction. (E) is the reduction half reaction taking place in given reaction.

Table 1.14 Oxidation Reduction (Redox) Reaction
Oxidation–reduction (or redox) reactions, are a type of chemical reaction that involves transfer of electron(s) between two species. In a redox reaction, the oxidation numbers (oxidation states) of the atoms are

changed. Oxidation refers to the loss of electrons (oxidation number increases), while reduction refers to the gain of electrons (oxidation number decreases).

For example, in the reaction of combustion of magnesium (Mg) in air, two electrons are transferred from Mg to O to form MgO:

$$Mg + O_2 \rightarrow MgO$$

This reaction may be written as two half–reaction.

$$Mg \rightarrow Mg^{2+} + e^- \text{ (oxidation reaction)}$$
$$O_2 + e^- \rightarrow O^{2-} \text{ (reduction reaction)}$$

In the above reaction, electrons are transferred from Mg to O_2. Mg is oxidized and is called a reducing reagent, and O_2 is reduced and is called an oxidizing reagent.

Similarly, in the reaction given:

$$Cr_2O_7^{2-} + 6Cl^- + 14H^+ \rightarrow 2Cr^{3+} + 7H_2O(l) + 3Cl_2(g)$$

Cl^- (oxidation number –1) is converted to Cl_2 (oxidation number 0), Cl^- must loses electron, and it's oxidized;

$$6Cl^- \rightarrow 3Cl_2(g) + 6e^- \quad \textbf{(oxidation half reaction)}$$

Cr in $Cr_2O_7^{2-}$ (oxidation number +6) is converted to Cr^{3+} (oxidation number +3), $Cr_2O_7^{2-}$ must gains electrons, and it's reduced. The reduction half–reaction is:

$$Cr_2O_7^{2-} + 14H^+ + 6e^- \rightarrow Cr^{3+} + 7H_2O(l) \quad \textbf{(reduction half reaction)}$$

57. **(D)** is correct answer. To titrate 100.0 mL of 1.0 M NaOH, the volume of 5.0 M of HCl needed is:

100.0 (mL) x 1.0 (M) / 5.0 (M) = 20 mL. Since the initial reading of buret for HCl is 3.15 mL, the final reading of HCl is 20 mL + 3.15 mL = 23.15 mL.

58. **(D)** is correct answer. Generally, solubility of a gas increase with increase in pressure (Henry's Law), and decrease with increase in temperature. Therefore, answers (A) and (B) are incorrect. When lip of the tube is lifted to level I, pressure of the gas in tube decreases, and solubility of the gas decreases, answer (C) is incorrect. In contrast, when lip of the tube is lifted to level II, pressure of the gas in tube increases, and solubility of the gas increases (**answer D is correct**). Answer (E) is incorrect since when the concentration of gas in tube decrease, the solubility also decrease.

59. **(D)** is correct answer. H_2 is a colorless gas and can be produced in the reaction between a metal and HCl. (A) is incorrect because Cl_2 is not colorless. (B) is incorrect since there is no reaction for HCl to produce O_2. (D) is incorrect since HCl must react with carbonate (or bicarbonate), rather than metal to form CO_2. (E) is incorrect since there is no such reaction for HCl to reaction with metal to produce NO_2.

60. **(D)** is correct answer since CH_4 is a nonpolar molecule although the C–H bond is polar. Therefore, the primary intermolecular force is dispersion (van der Waals) force.

61. **(B)** is correct answer. When a hydrated salt sample (136 g) is heated, its water is lost, and the nonhydrated salt remains (100 g). The mass of hydrate water is 136 g – 100 g = 36 g. Therefore, the percent of water is calculated as: $\frac{36}{136} \times 100$.

62. **(B)** The balanced equation is: __1__ C_3H_8 + __5__ O_2 → __3__ CO_2 + __4__ H_2O

63. **(D)** is correct answer. Half–life ($t_{1/2}$) of a radioactive isotope is the amount of time required for half of it to decay as measured at the beginning of the time period. In this question, the half life of ^{112}Ag is 3.13 hours, after 6.26 hour, i.e. 2 half–lives, the amount of original ^{112}Ag will be halved twice, and there will be ¼ of original ^{112}Ag left, i.e. 25 g.

64. **(D)** is the correct answer. In a periodic table, first ionization energy increases from left to right in a row, and decreases down a column (**Figure** 1.3). Fluorine is located at right upper corner of the periodic table, and is one of the elements which have highest first ionization energy (only two elements, He and Ne, have higher first ionization energies than F).

65. **(C)** The strength of an acid refers to its ability or tendency to lose (donate) a proton (H^+). A strong acid, such as HCl, HNO_3, is one that completely ionizes (dissociates) in a solution. For a weak acid, the ability to donate proton (H^+) varies, and it's expressed as K_a, the greater the K_a, the greater its tendency to lose a proton. Therefore, the weak acid with greatest K_a is the strongest acid.

66. **(E)** is correct answer. All acid/base need to be properly disposed, direct pouring them into sink is not allowed (I is incorrect). Never smell or taste any chemicals in the lab (II is incorrect). When heating a substance in a test tube, leave the tube open to prevent buildup of pressure and cause explosion (III is incorrect).

67. **(C)** Diamond is an example of network covalence solid (**Table** 1.2). Other examples of network covalent bond: graphite (C), quartz (SiO_2).

68. **(E)** is correct answer. Average kinetic energy increase with temperature, not inversely related.

    Kinetic molecular theory is a theoretical model, describing behavior of gas based on the following postulates, or assumptions.

    a) Gases are composed of a large number of particles that behave like hard, spherical objects in a state of constant, random motion (D).
    b) Gas particles move in a straight line until they collide with other particles or the walls of the container.
    c) Gas particles are much smaller than the distance between particles. Most of the volume of a gas is empty space (A).
    d) There is no force of attraction between gas particles or between the particles and the walls of the container (B).
    e) Collisions between gas particles or collisions with the walls of the container are perfectly elastic. Energy of a gas particle is not lost when it collides with another particle or with the walls of the container (C).
    f) The average kinetic energy of a collection of gas particles depends on the temperature of the gas (A).

69. **(C)** Elements in the same group have the same number of valence electrons (outer shell configuration), and therefore, they have similar chemical properties.

70. **(D)** is a substitution (single displacement) reaction, not an addition reaction. Other reactions are classified correctly.

Table 1.15 *Common Types of Reaction*		
Type	**General Equation**	**Notes**
Synthesis (combination)	$A + B \rightarrow AB$	Two or more reactants combine to form one product
Decomposition	$AB \rightarrow A + B$	One reactant to split to two or more products
Single displacement (substitution)	$A + BC \rightarrow AC + B$	Part of a compound is replaced by different element (group)
Double displacement	$AB + CD \rightarrow AC + BD$	Two compound exchange part of the compound
Neutralization (acid/base)	$HA + BOH \rightarrow BA + H_2O$	Acid and base react to produce a salt and water
Combustion	$A + O_2 \rightarrow$ one or more oxides	Reducing reactant vigorously react with O_2 to generate flame.

101. **Correct answer: I True, II True, CE No**

    Explanation: Both statements are true, but statement II is not the correct explanation of statement I. The color of potassium permanganate is caused by presence of manganese which is a transition metal. Compound with transition metal has color due to the unfilled d orbitals.

102. **Correct answer: I True, II True, CE No**

    Explanation: Both statements are true, but statement II does not explain why water molecule is polar. H_2O molecule is polar because H–O bond is polar and the fact that the two O–H bonds are not linear (statement I is correct). Combustion of hydrogen release large amount of heat, and is exothermic reaction (statement II is correct).

103. **Correct answer: I True, II True, CE Yes**

    Isotopes of the same element have the same number of proton and electron, and subsequently the same electron configuration and valence electron configuration. The chemical properties of elements are determined by the valance electron configuration. Both statements are correct and statement II is correct explanation for statement I.

104. **Correct answer: I False, II True, CE No**

    Explanation: Boyle's law states that the volume of a gas is in reverse proportion with pressure when temperature is constant. Therefore, statement I is incorrect. Statement II is correct definition of Boyle's law, but not the correct explanation of statement I because statement I is incorrect.

105. **Correct answer: I True, II True, CE Yes**

    Explanation: Both statements are true and statement II is correct explanation of statement I, because the boiling point of a liquid is determined by the surrounding air. When the vapor pressure of a liquid equals to the surrounding air, the liquid starts boil, and the vapor pressure of a liquid is determined by temperature. Therefore, the boiling point of a liquid varies with pressure of surrounding air.

106. **Correct answer: I True, II True, CE Yes**

    Explanation: Both statements are true. Chlorine has two isotopes, ^{35}Cl and ^{37}Cl, with abundance of 75.75% and 24.25% respectively. The average of atomic mass of chlorine is 35.45, which is calculated from the atomic masses of the two isotopes and their abundance as: 35 x 76.75% + 37 x 24.25% = 35.45.

107. **Correct answer: I True, II True, CE No**

    Explanation: An element that has the electron configuration $1s^2 2s^2 2p^6 3s^2 3p^6 3d^3 4s^2$ has completely filled $1s$, $2s$, $2p$, $3s$, and $3p$ orbitals and partially filled $3d$ orbitals, therefore, both statements I and II are correct. However, the reason for this element to be a transition metal is because of its partially filled $3d$ orbitals, not its completely filled $1s$, $2s$, $2p$, $3s$, and $3p$ orbitals. In fact, all elements heavier than argon (mass number 18) have completely filled $1s$, $2s$, $2p$, $3s$, and $3p$ orbitals.

108. **Correct answer: I True, II False, CE No**

    Explanation: Within a group, both the radius of atom and atomic number increase downward, the higher the atomic number, the bigger the radius. Therefore, statement I is correct, but statement II is incorrect. This is different from the elements in a row (period), in which radius decrease with increase in atomic number.

109. **Correct answer: I True, II True, CE Yes**

    Explanation: Both statements are true, and statement II correctly explains statement I. *Nonpolar covalence bond* is result of equal sharing of electron pair between two atoms with the same *electronegativity*. Such bonds are formed between two *nonmetallic atoms* of the same elements, such as H_2, O_2, N_2, F_2, Cl_2, Br_2, I_2.

110. **Correct answer: I True, II True, CE Yes**

Explanation: Both statements are true, and statement II correctly explains statement I. *Empirical formula* of a chemical compound is the simplest positive integer ratio of atoms present in a compound. According to this definition, CH_2O is the correct empirical formula of $C_6H_{12}O_6$.

111. **Correct answer: I False, II true, CE No**

   Explanation: Statement I is false since a solution of pH = 5 is **acidic**, not basic. Statement II is correct, since concentrations of H^+ and OH^- in solution of pH = 5 is 10^{-5} M and 10^{-9} M respectively.

112. **Correct answer: I True, II False, CE No**

   Explanation: Increase reactants concentration increase collision between reactant molecules, and therefore, increase reaction rate (statement I is true). But, higher concentration of reactant has no impact on activation energy (statement II is false). Catalyst can lower activation energy.

113. **Correct answer: I True, II False, CE No**

   Explanation: Statement I is true, but statement II is false. The correct statement is "a conjugate base is formed once a Brønsted –Lowry acid **loses** a proton, not gains a proton.

114. **Correct answer: I False, II True, CE No**

   Explanation: Statement I is false, the right half reaction is $F_2 + 2e^- \rightarrow 2F^-$. Like full reaction equation, half reaction must satisfy conservation of mass and charge (Statement II is true).

115. **Correct answer: I True, II False, CE No**

   Explanation: ethane (CH_3–CH_3) is a saturated hydrocarbon (statement I is correct), but ethane (CH_2=CH_2) has double bond between two C atoms, not triple bond (statement II is false).

SAT II Chemistry

Practice Test 2

Answer Sheet

Part A and C: Determine the correct answer. Blacken the oval of your choice completely with a No. 2 pencil.

1	Ⓐ Ⓑ Ⓒ Ⓓ Ⓔ	25	Ⓐ Ⓑ Ⓒ Ⓓ Ⓔ	49	Ⓐ Ⓑ Ⓒ Ⓓ Ⓔ
2	Ⓐ Ⓑ Ⓒ Ⓓ Ⓔ	26	Ⓐ Ⓑ Ⓒ Ⓓ Ⓔ	50	Ⓐ Ⓑ Ⓒ Ⓓ Ⓔ
3	Ⓐ Ⓑ Ⓒ Ⓓ Ⓔ	27	Ⓐ Ⓑ Ⓒ Ⓓ Ⓔ	51	Ⓐ Ⓑ Ⓒ Ⓓ Ⓔ
4	Ⓐ Ⓑ Ⓒ Ⓓ Ⓔ	28	Ⓐ Ⓑ Ⓒ Ⓓ Ⓔ	52	Ⓐ Ⓑ Ⓒ Ⓓ Ⓔ
5	Ⓐ Ⓑ Ⓒ Ⓓ Ⓔ	29	Ⓐ Ⓑ Ⓒ Ⓓ Ⓔ	53	Ⓐ Ⓑ Ⓒ Ⓓ Ⓔ
6	Ⓐ Ⓑ Ⓒ Ⓓ Ⓔ	30	Ⓐ Ⓑ Ⓒ Ⓓ Ⓔ	54	Ⓐ Ⓑ Ⓒ Ⓓ Ⓔ
7	Ⓐ Ⓑ Ⓒ Ⓓ Ⓔ	31	Ⓐ Ⓑ Ⓒ Ⓓ Ⓔ	55	Ⓐ Ⓑ Ⓒ Ⓓ Ⓔ
8	Ⓐ Ⓑ Ⓒ Ⓓ Ⓔ	32	Ⓐ Ⓑ Ⓒ Ⓓ Ⓔ	56	Ⓐ Ⓑ Ⓒ Ⓓ Ⓔ
9	Ⓐ Ⓑ Ⓒ Ⓓ Ⓔ	33	Ⓐ Ⓑ Ⓒ Ⓓ Ⓔ	57	Ⓐ Ⓑ Ⓒ Ⓓ Ⓔ
10	Ⓐ Ⓑ Ⓒ Ⓓ Ⓔ	34	Ⓐ Ⓑ Ⓒ Ⓓ Ⓔ	58	Ⓐ Ⓑ Ⓒ Ⓓ Ⓔ
11	Ⓐ Ⓑ Ⓒ Ⓓ Ⓔ	35	Ⓐ Ⓑ Ⓒ Ⓓ Ⓔ	59	Ⓐ Ⓑ Ⓒ Ⓓ Ⓔ
12	Ⓐ Ⓑ Ⓒ Ⓓ Ⓔ	36	Ⓐ Ⓑ Ⓒ Ⓓ Ⓔ	60	Ⓐ Ⓑ Ⓒ Ⓓ Ⓔ
13	Ⓐ Ⓑ Ⓒ Ⓓ Ⓔ	37	Ⓐ Ⓑ Ⓒ Ⓓ Ⓔ	61	Ⓐ Ⓑ Ⓒ Ⓓ Ⓔ
14	Ⓐ Ⓑ Ⓒ Ⓓ Ⓔ	38	Ⓐ Ⓑ Ⓒ Ⓓ Ⓔ	62	Ⓐ Ⓑ Ⓒ Ⓓ Ⓔ
15	Ⓐ Ⓑ Ⓒ Ⓓ Ⓔ	39	Ⓐ Ⓑ Ⓒ Ⓓ Ⓔ	63	Ⓐ Ⓑ Ⓒ Ⓓ Ⓔ
16	Ⓐ Ⓑ Ⓒ Ⓓ Ⓔ	40	Ⓐ Ⓑ Ⓒ Ⓓ Ⓔ	64	Ⓐ Ⓑ Ⓒ Ⓓ Ⓔ
17	Ⓐ Ⓑ Ⓒ Ⓓ Ⓔ	41	Ⓐ Ⓑ Ⓒ Ⓓ Ⓔ	65	Ⓐ Ⓑ Ⓒ Ⓓ Ⓔ
18	Ⓐ Ⓑ Ⓒ Ⓓ Ⓔ	42	Ⓐ Ⓑ Ⓒ Ⓓ Ⓔ	66	Ⓐ Ⓑ Ⓒ Ⓓ Ⓔ
19	Ⓐ Ⓑ Ⓒ Ⓓ Ⓔ	43	Ⓐ Ⓑ Ⓒ Ⓓ Ⓔ	67	Ⓐ Ⓑ Ⓒ Ⓓ Ⓔ
20	Ⓐ Ⓑ Ⓒ Ⓓ Ⓔ	44	Ⓐ Ⓑ Ⓒ Ⓓ Ⓔ	68	Ⓐ Ⓑ Ⓒ Ⓓ Ⓔ
21	Ⓐ Ⓑ Ⓒ Ⓓ Ⓔ	45	Ⓐ Ⓑ Ⓒ Ⓓ Ⓔ	69	Ⓐ Ⓑ Ⓒ Ⓓ Ⓔ
22	Ⓐ Ⓑ Ⓒ Ⓓ Ⓔ	46	Ⓐ Ⓑ Ⓒ Ⓓ Ⓔ	70	Ⓐ Ⓑ Ⓒ Ⓓ Ⓔ
23	Ⓐ Ⓑ Ⓒ Ⓓ Ⓔ	47	Ⓐ Ⓑ Ⓒ Ⓓ Ⓔ	71	Ⓐ Ⓑ Ⓒ Ⓓ Ⓔ
24	Ⓐ Ⓑ Ⓒ Ⓓ Ⓔ	48	Ⓐ Ⓑ Ⓒ Ⓓ Ⓔ	72	Ⓐ Ⓑ Ⓒ Ⓓ Ⓔ

Part B: On the actual Chemistry Test, the following type of question must be answered on a special section (labeled "Chemistry") at the lower left–hand corner of your answer sheet. These questions will be numbered beginning with 101 and must be answered according to the directions.

PART B	I		II		CE
101	T	F	T	F	◯
102	T	F	T	F	◯
103	T	F	T	F	◯
104	T	F	T	F	◯
105	T	F	T	F	◯
106	T	F	T	F	◯
107	T	F	T	F	◯
108	T	F	T	F	◯
109	T	F	T	F	◯
110	T	F	T	F	◯
111	T	F	T	F	◯
112	T	F	T	F	◯
113	T	F	T	F	◯
114	T	F	T	F	◯
115	T	F	T	F	◯
116	T	F	T	F	◯

Periodic Table of Elements

Material in this table may be useful in answering the questions in this examination

1																	2
H 1.0079																	**He** 4.0026
3 **Li** 6.941	4 **Be** 9.012											5 **B** 10.811	6 **C** 12.011	7 **N** 14.007	8 **O** 16.00	9 **F** 19.00	10 **Ne** 20.179
11 **Na** 22.99	12 **Mg** 24.30											13 **Al** 26.98	14 **Si** 28.09	15 **P** 30.974	16 **S** 32.06	17 **Cl** 35.453	18 **Ar** 39.948
19 **K** 39.01	20 **Ca** 40.48	21 **Sc** 44.96	22 **Ti** 47.90	23 **V** 50.94	24 **Cr** 52.00	25 **Mn** 54.938	26 **Fe** 55.85	27 **Co** 58.93	28 **Ni** 58.69	29 **Cu** 63.55	30 **Zn** 65.39	31 **Ga** 69.72	32 **Ge** 72.59	33 **As** 74.92	34 **Se** 78.96	35 **Br** 79.90	36 **Kr** 83.80
37 **Rb** 85.47	38 **Sr** 87.62	39 **Y** 88.91	40 **Zr** 91.22	41 **Nb** 92.91	42 **Mo** 95.94	43 **Tc** (98)	44 **Ru** 101.1	45 **Rh** 102.91	46 **Pd** 106.42	47 **Ag** 107.87	48 **Cd** 112.41	49 **In** 114.82	50 **Sn** 118.71	51 **Sb** 121.75	52 **Te** 127.60	53 **I** 126.91	54 **Xe** 131.29
55 **Cs** 132.91	56 **Ba** 137.33	57 ***La** 138.91	72 **Hf** 178.49	73 **Ta** 180.95	74 **W** 183.85	75 **Re** 186.21	76 **Os** 190.2	77 **Ir** 192.2	78 **Pt** 195.08	79 **Au** 196.97	80 **Hg** 200.59	81 **Tl** 204.38	82 **Pb** 207.2	83 **Bi** 208.98	84 **Po** (209)	85 **At** (210)	86 **Rn** (222)
87 **Fr** (223)	88 **Ra** 226.02	89 **Ac** 227.03	104 **Rf** (261)	105 **Db** (262)	106 **Sg** (266)	107 **Bh** (264)	108 **Hs** (277)	109 **Mt** (268)	110 **Ds** (271)	111 **Rg** (272)	112 (277)						

	58	59	60	61	62	63	64	65	66	67	68	69	70	71
*Lanthanide Series	**Ce** 140.12	**Pr** 140.91	**Nd** 144.24	**Pm** (145)	**Sm** 150.4	**Eu** 151.97	**Gd** 157.25	**Tb** 158.93	**Dy** 162.50	**Ho** 164.93	**Er** 167.26	**Tm** 168.93	**Yb** 173.04	**Lu** 174.97
	90	91	92	93	94	95	96	97	98	99	100	101	102	103
Actinide Series	**Th** 232.04	**Pa** 231.04	**U** 238.03	**Np** 237.05	**Pu** (244)	**Am** (243)	**Cm** (247)	**Bk** (247)	**Cf** (251)	**Es** (252)	**Fm** (257)	**Md** (258)	**No** (259)	**Lr** (260)

Note: For all questions involving solutions, assume that the solvent is water unless otherwise stated.
Reminder: You may not use a calculator in this test!

Throughout the test the following symbols have the definitions specified unless otherwise noted.

H = enthalpy	atm = atmosphere(s)
M = molar	g = gram(s)
n = number of moles	J = joule(s)
P = pressure	kJ = kilojoule(s)
R = molar gas constant	L = liter(s)
S = entropy	mL = milliliter(s)
T = temperature	mm = millimeter(s)
V = volume	mol = mole(s)
	V = volt(s)

Chemistry Subject Practice Test 2

Part A

Directions for Classification Questions
Each set of lettered choices below refers to the numbered statements or questions immediately following it. Select the one lettered choice that best fits each statement or answers each question and then fill in the corresponding circle on the answer sheet. A choice may be used once, more than once, or not at all in each set.

Questions 1 – 5 refer to the following electron configuration

    (A) $1s^2 2s^2 2p^6 3p^1$
    (B) $1s^2 2s^2 2p^6 3s^2 3p^6$
    (C) $1s^2 2s^2 2p^6 3s^2 3p^6 4s^2$
    (D) $1s^2 2s^2 2p^6 3s^2 3p^6 4s^1$
    (E) $1s^2 2s^2 2p^6 3s^2 3p^6 3d^6 4s^2$

1. Which is the electron configuration of calcium ion, Ca^{2+}.

2. Which is the electron configuration of a transition metal.

3. Which is the electron configuration of an excited atom.

4. Which is the electron configuration of potassium in the ground state.

5. Which is the electron configuration of a noble gas.

Questions 6 – 8 refer to following organic structural formula

    (A) R–OH
    (B) R–O–R'
    (C) R–COOH
    (D) R–COO–R'
    (E) R–CO–R'

6. Which represents an ester.

7. Which represents a ketone.

8. Which represents a carboxylic acid.

Questions 9 – 12 refer to following pairs of substance

    (A) Br_2 and Hg
    (B) Cl_2 and F_2
    (C) NH_4^+ and H_3O^+
    (D) Fe and Co
    (E) Diamond and graphite

9. Which are liquids at room temperature (293 K).

10. Which have coordinate covalent bonds.

11. Which are allotropes of each other.

12. Which are in the same group on the periodic table, and are both strong oxidizing reagents.

Questions 13 – 15 refer to the following

    (A) Arrhenius acid
    (B) Arrhenius base
    (C) Electrolyte
    (D) Indicator
    (E) Salt

13. Has different colors in its acid and basic forms

14. At 25°C, produces a aqueous solution with pH < 7

15. At 25°C, produces a aqueous solution with $[OH^-] < 1.0 \times 10^{-7}$ moles per liter

GO ON TO THE NEXT PAGE

Questions 16 – 19 refer to the following diagrams

(A)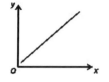

(B)

(C)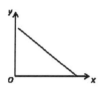

(D)

(E)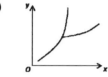

16. Could be a diagram showing different states of a substance at different temperature (x) and pressure (y).

17. Could be a plot of the pressure (y) against the absolute temperature (x) for an ideal gas in a fixed volume — Charles' Law.

18. Could be the plot of an average molecule kinetic energy of molecules (y) against the absolute temperature (x) for an ideal gas.

19. Could be a plot of the pressure (y) against the volume (x) for an ideal gas at fixed temperature — Boyle's Law.

Questions 20 – 22 refer to the following nuclear reactions

    (A) Gamma decay
    (B) Beta decay
    (C) Alpha decay
    (D) Nuclear fission
    (E) Nuclear fusion

20. The process through which hydrogen atoms are merged to produce helium

21. The process that results in no change in the mass number, but increase in atomic number by 1

22. The nuclear process that transmutes uranium–238 into thorium–234

Questions 23 – 25 refer to the following numbers

    (A) 1
    (B) 2
    (C) 3
    (D) 4
    (E) 5

The coefficient of a compound in reaction when the equation is balanced and all the coefficients are reduced to the simplest whole number.

23. H_2SO_4 in $HMnO_4 + H_2SO_3 \rightarrow MnSO_4 + H_2SO_4 + H_2O$

24. CO_2 in $Fe_2O_3 + CO \rightarrow Fe + CO_2$

25. CO_2 in $C_2H_4 + O_2 \rightarrow CO_2 + H_2O$

GO ON TO THE NEXT PAGE

PLEASE GO TO THE SPECIAL SECTION AT THE LOWER LEFT–HAND CORNER OF PAGE 2 OF YOUR ANSWER SHEET LABELED CHEMISTRY AND ANSWER QUESTIONS 101–115 ACCORDING TO THE FOLLOWING DIRECTIONS.

Part B

Directions for Relationship Analysis Questions

Each question below consists of two statements, I in the left–hand column and II in the right–hand column. For each question, determine whether statement I is true or false and whether statement II is true or false and fill in the corresponding T or F circles on your answer sheet. *Fill in circle CE only if statement II is a correct explanation of the true statement I.*

EXAMPLES:

I		II
EX1. The nucleus in an atom has a positive charge.	BECAUSE	Proton has positive charge, neutron has no charge.

		I	II	CE
SAMPLE ANSWERS	EX1	● Ⓕ	● Ⓕ	●

	I		II
101.	The reaction of hydrogen with oxygen to form water is an exothermic reaction	BECAUSE	the bonds in a water molecule are polar covalent bonds.
102.	The element with an electron configuration of $[He]2s^2$ has a larger atomic radius than fluorine	BECAUSE	the element with an electron configuration of $[He]2s^2$ has a greater nucleus charge than fluorine.
103.	1 M $NaCl(aq)$ will have a higher boiling point than that of 1 M $CaCl_2(aq)$	BECAUSE	1 mole of NaCl yields 1 moles less of ions in solution than that 1 moles of $CaCl_2$ yields.
104.	The solubility of gases in water does not depend on pressure	BECAUSE	the vapor pressure of a liquid does not change with surrounding atmosphere at constant temperature.
105.	After a system has reached chemical equilibrium, there is no change in the concentrations of reactants and products	BECAUSE	reactions are not reversible at equilibrium.

GO ON TO THE NEXT PAGE

106. Powdered limestone ($CaCO_2$) reacts faster with acid than a larger piece of limestone   BECAUSE   powdered limestone has a greater surface area.

107. HCl cannot be collected by water displacement   BECAUSE   H–Cl is a nonpolar bond.

108. Pure water can boil at a temperature less than 100°C   BECAUSE   water boils when the vapor pressure of the water is equal to the atmospheric pressure.

109. An endothermic reaction has a positive value for ΔH BECAUSE   in an endothermic reaction, the products have less potential energy than the reactants.

110. As pressure on a gas increases, the volume of the gas decreases   BECAUSE   when temperature of a gas increases, its volume increases.

111. The addition of H_2 to ethene will form an unsaturated compound called ethane   BECAUSE   An ethane molecule has a double bond between two C atoms.

112. AgCl is insoluble in water   BECAUSE   the reaction $Ag^+(aq) + Cl^-(aq) \rightarrow AgCl(s)$ is not reversible.

113. Catalyst decrease the rate of chemical reaction   BECAUSE   catalyst decrease the activation energy.

114. Electrolysis of water requires input of energy   BECAUSE H_2 and O_2 have higher chemical potential energy than H_2O.

115. CH_3CH_2-OH and CH_3-O-CH_3 are isomers   BECAUSE CH_3CH_2-OH and CH_3-O-CH_3 have the same molecular formula.

GO ON TO THE NEXT PAGE

Part C

Directions for Five–Choice Completion Questions
Each of the questions or incomplete statements below is followed by five suggested answers or completions. Select the one that is best in each case and then fill in the corresponding circle on the answer sheet.

26. If 10. 0 mL of 1.0 M HCl was required to titrate a 50.0 mL NaOH solution of unknown concentration to its end point, what was the concentration of the NaOH?

(A) 0.1 M
(B) 0.2 M
(C) 0.5 M
(D) 2.0 M
(E) 5.0 M

27. Which pair of numbers below is correct regarding the number of pi bond and number of atoms lying in a straight line in the molecule propyne, $HC \equiv C–CH_3$?

(A) 1 pi bond and 3 atoms in a straight line
(B) 2 pi bonds and 3 atoms in a straight line
(C) 2 pi bonds and 4 atoms in a straight line
(D) 4 pi bonds and 4 atoms in a straight line
(E) 6 pi bonds and 6 atoms in a straight line

28. Which of the following conditions guarantee a spontaneous reaction?

(A) positive ΔH, positive ΔS
(B) positive ΔH, negative ΔS
(C) negative ΔH, negative ΔS
(D) negative ΔH, positive ΔS
(E) none of the above

29. What is the concentration of hydroxide ion, OH^-, in a solution with a pH of 5?

(A) 1.0×10^{-3}
(B) 1.0×10^{-5}
(C) 1.0×10^{-7}
(D) 1.0×10^{-9}
(E) 1.0×10^{-11}

30. The reaction of zinc metal and HCl produces which of the following?

I. $H_2(g)$
II. $Cl_2(g)$
III. $ZnCl_2(aq)$

(A) I only
(B) III only
(C) I and II only
(D) I and III only
(E) I, II, and III

31. For a particular compound, which of the following pairs correctly represent the empirical and the molecular formula, respectively?

(A) C_2H_3 and C_4H_{10}
(B) CH and C_6H_6
(C) CH_2 and C_3H_8
(D) CH and C_2H_4
(E) CH_3 and CH_4

32. Which of the following CANNOT be concluded based on observations of Rutherford's gold foil experiment?

(A) Most alpha particles can pass through a thin sheet of gold foil
(B) Some alpha particles are deflected or reflected
(C) The atom is mostly empty space
(D) The nucleus has a negative charge
(E) The atom has a small but dense center

GO ON TO THE NEXT PAGE

X → Y

33. In a reaction represented above, the potential energy of the X is 35 kJ/mol, the potential energy of Y is 15 kJ/mol and the potential energy of the activated complex is 52 kJ/mol. What is the activation energy for the reverse reaction?

(A) 20 kJ/mol
(B) –20 kJ/mol
(C) 17 kJ/mol
(D) 37 kJ/mol
(E) 67 kJ/mol

34. Which of the following reactions would form at least one solid precipitate as a product?

   I. $AgNO_3(aq) + KCl(aq) \rightarrow$
  II. $Pb(NO_3)_2(aq) + 2NaBr(aq) \rightarrow$
 III. $Ba(OH)_2(aq) + Li_2SO_4(aq) \rightarrow$

(A) I only
(B) II only
(C) III only
(D) I and II only
(E) I, II, and III

$A(aq) + B(aq) \rightleftharpoons C(aq) + D(s)$

35. The equilibrium constant for the reaction above is given by the expression

(A) $\dfrac{[C][D]}{[A][B]}$

(B) $\dfrac{[C]+[D]}{[A]+[B]}$

(C) $\dfrac{[C][D]}{[A]+[B]}$

(D) $\dfrac{[C]}{[A][B]}$

(E) $\dfrac{[D]}{[A][B]}$

36. Which of the following statements about catalysts is true?

(A) They increase the value of the equilibrium constant.
(B) They reduce the concentration of product.
(C) They increase the concentration of reactants.
(D) They increase the potential energy of reactants.
(E) They reduce the activation energy of the reaction.

37. $H_2O(l) + S^{2-}(aq) \rightleftharpoons HS^-(aq) + OH^-(aq)$

In the equation for the reaction represented above, the species acting as acids are

(A) $H_2O(l)$ and S^{2-}
(B) $H_2O(l)$ and HS^-
(C) $H_2O(l)$ and OH^-
(D) S^{2-} and HS^-
(E) S^{2-} and OH^-

38. How many grams of $Pb(NO_3)_2$ (formula mass = 331 g/mol) is needed to make 100 mL solution containing 1.0 M of NO_3^- ion.

(A) 8.3
(B) 16.5
(C) 33.1
(D) 66.2
(E) 165

39. Characteristic features of naturally occurring radioactive elements include which of the following?

   I.   Spontaneous decay
  II.   A characteristic half–life
 III.   The emission of α or β particles or γ rays

(A) I only
(B) II only
(C) I and II only
(D) II and III only
(E) I, II and III

GO ON TO THE NEXT PAGE

40. Which aqueous solution is expected to have the highest boiling point under the same pressure?

    (A) 0.1 mole/L $AlCl_3$
    (B) 0.2 mole/L $CaCl_2$
    (C) 0.4 mole/L NaCl
    (D) 0.5 mole/L CH_3O
    (E) They boil at the same temperature

41. In which of the following process entropy decreases?

    (A) Dissolving a salt into a solution
    (B) Liquid water is evaporated
    (C) Burning a log in a fireplace
    (D) Formation of dry ice from CO_2 gas at low temperature
    (E) Ice in a cup melts

42. $...ClO^- \rightleftharpoons ...ClO_3^- + ...Cl^-$

    When the equation for the reaction represented above is balanced with coefficients reduced to the lowest whole-number terms, which of the following statements is correct?

    I.   The coefficient for Cl^- is 4
    II. The coefficient for ClO^- is 3
    III. The coefficient for ClO_3^- is 2

    (A) I only
    (B) II only
    (C) I and II only
    (D) II and III only
    (E) I, II and III

43. The following practices are considered to be safe in a chemistry laboratory EXCEPT?

    (A) Using a test tube holder to handle a hot test tube
    (B) Tying long hair back before experiment
    (C) Eating lunch on carefully cleaned bench top
    (D) Pouring liquids while holding the reagent bottles over the sink
    (E) Working under a fume hood

44. A 5–mole sample of Ar gas is added to a container with 1 mole H_2 gas in it. Without changing temperature and the volume of the container, which of the following statements regarding the partial pressure of H_2 exerted on the wall of the container is true after the two gases uniformly mixed?

    (A) One fifth of the original pressure
    (B) The same as the original pressure
    (C) Twice the original pressure
    (D) Five times the original pressure
    (E) Six times the original pressure

45. For the reaction $Mg(s) + O_2(g) \rightarrow MgO(s)$. If 48.6 grams of Mg strip is placed in a closed container with 64.0 grams of O_2. The reaction is allowed to proceed to completion. How much MgO is produced?

    (A) 15.4 grams
    (B) 40.3 grams
    (C) 80.6 grams
    (D) 96.2 grams
    (E) 112.6 grams

46. At 23°C, 200 mililiters of an ideal gas exerts a pressure of 750 milimeters of mercury. The volume of the gas at 0°C and 760 milimeters of mercury is calculated from which of the following expression?

    (A) $200 \times \frac{760}{750} \times \frac{273}{296}$ mL

    (B) $200 \times \frac{750}{760} \times \frac{0}{23}$ mL

    (C) $200 \times \frac{760}{750} \times \frac{23}{0}$ mL

    (D) $200 \times \frac{760}{750} \times \frac{296}{273}$ mL

    (E) $200 \times \frac{750}{760} \times \frac{273}{296}$ mL

GO ON TO THE NEXT PAGE

47. Given the reaction at equilibrium in a closed container:

A + B ⇌ AB + heat

Which of the following conditions will shift the reaction so that the formation of AB is favored?

   I. Removing AB from the system
  II. Increase the temperature
 III. Adding more reactant A

(A) I only
(B) III only
(C) I and II only
(D) I and III only
(E) I, II, and III

48. Which of the following has the lowest electronegativity?

(A) Ca
(B) Cl
(C) K
(D) Si
(E) Fe

49. Given the same condition of temperature and pressure, which gas below has the greatest rate of effusion?

(A) H_2
(B) Ar
(C) O_2
(D) F_2
(E) Cl_2

50. Which of the following statement about ideal gas particles is correct?

(A) Have no attraction between them
(B) Have sizes similar to those of H atoms
(C) Travel with a straight line motion
(D) Never collide with each other
(E) One mole of them always occupy 22.4 L of volume

51. Which of the following atoms has largest radius?

(A) Na
(B) S
(C) Cs
(D) Al
(E) Cl

52. Which of the following represents the geometric shape of a molecule with sp^2 hybridization has?

(A) Linear
(B) Trigonal pyramid
(C) Trigonal planar
(D) Square
(E) Tetrahedral

53. $Cl_2(g) + KBr(aq) \rightarrow$

When 1 mole of chlorine gas reacts completely with excess KBr solution, the products obtained are

(A) 1 mole of Cl^- ions and 1 mole of Br
(B) 1 mole of Cl^- ions and 2 mole of Br
(C) 1 mole of Cl^- ions and 1 mole of Br_2
(D) 2 mole of Cl^- ions and 1 mole of Br_2
(E) 2 mole of Cl^- ions and 2 mole of Br_2

54. $...AlCl_3(aq) + ...NH_3(aq) + ...H_2O \rightarrow$

Which of the following is one of the products obtained from the reaction above?

(A) AlN_3
(B) AlH_3
(C) $Al(NH_3)_3$
(D) $Al(NO_3)_3$
(E) $Al(OH)_3$

GO ON TO THE NEXT PAGE

55. A neutral atom has a total of 17 electrons. The electron configuration in the outermost principle energy level is

(A) $3s^5 3p^2$
(B) $3s^2 3p^5$
(C) $3s^2 3p^5$
(D) $3s^2 3p^6 3d^7$
(E) $3s^2 3p^7$

56. Which of the following combinations of particles represents an ion of net charge of +1 and of mass number of 85?

(A) 47 neutrons, 38 protons, 36 electrons
(B) 47 neutrons, 37 protons, 36 electrons
(C) 47 neutrons, 38 protons, 38 electrons
(D) 48 neutrons, 37 protons, 36 electrons
(E) 48 neutrons, 36 protons, 37 electrons

57. Which of the following is correct about the subatomic particles found in $^{37}Cl^{-1}$?

  I.   21 neutrons
  II. 17 protons
  III. 17 electrons

(A) I only
(B) II only
(C) III only
(D) I and II only
(E) II and III only

58. To determine the number of hydrate water (X) in $CuCl_2 \cdot XH_2O$, a student conducted an experiment in which he drove off all the hydrate water in the hydrated salt by heating, and weigh the sample before and after the heating process. With the result below, what is the value of X?

Mass of original salt before heating = 170.5 g
Mass of salt after heating = 134.5 g

(A) 1
(B) 2
(C) 3
(D) 4
(E) 5

59. The oxidation of state of sulfur is most positive in which of the following species?

(A) SO_4^{2-}
(B) SO_3^{2-}
(C) SO_2
(D) S_8
(E) H_2S

60. To prepare one molal of NaCl solution, one mole of sodium chloride is dissolved in which of the following amounts of water?

(A) One liter
(B) One hundred mililiters
(C) One hundred grams
(D) One kilogram
(E) One mole

61. What is the ΔH value for the reaction

$$N_2O_4 \rightarrow 2NO_2$$

given:

$2NO_2 \rightarrow N_2 + 2O_2 \quad \Delta H = -67.8$ kilojoules
$N_2 + 2O_2 \rightarrow N_2O_4 \quad \Delta H = +9.7$ kilojoules

(A) +58.1 kilojoules
(B) +77.5 kilojoules
(C) +116.2 kilojoules
(D) −58.1 kilojoules
(E) −77.5 kilojoules

62. What is the potential of the reaction below

$$2Fe^{2+} + Cl_2 \rightarrow 2Fe^{3+} + 2Cl^-$$

given the half–reaction potentials:

$Fe^{3+} + e^- \rightarrow Fe^{2+} \quad E° = +0.77$ V
$Cl_2 + 2e^- \rightarrow 2Cl^- \quad E° = +1.36$ V

(A) 0.18 V
(B) 0.59 V
(C) −0.59 V
(D) 2.13 V
(E) −0.18 V

GO ON TO THE NEXT PAGE

63. Which of the following aqueous solutions is the best electricity conductor?

(A) 1.0 M glucose, $C_6H_{12}O_6$
(B) 0.1 M hydrochloric acid, HCl
(C) 1.0 M sodium nitrate, $NaNO_3$
(D) 1.0 M acetic acid, $HC_2H_3O_2$
(E) 1.0 M methanol, CH_3OH

64. Which of the following represents a correctly balanced half–reaction?

(A) $Cl_2 + 2e^- \rightarrow Cl^-$
(B) $2e^- + Cu \rightarrow Cu^{2+}$
(C) $O_2 \rightarrow 2e^- + 2O^{2-}$
(D) $Al^{3+} + 3e^- \rightarrow Al$
(E) $2H^+ + e^- \rightarrow H_2$

65. What is the percent composition by mass of lead (Pb) in lead sulfate, $PbSO_4$ (formula mass = 303)?

(A) 35.4% Pb
(B) 44.7% Pb
(C) 56.9% Pb
(D) 68.3% Pb
(E) 87.5% Pb

66. Which of the following statements about solubility of gas in water is correct?

(A) Solubility of a gas in water increases when temperature increases
(B) Solubility of a gas in water increases when pressure increases
(C) Henry's law describes the relationship between solubility of a gas and temperature at constant pressure
(D) Solubility of a gas in water increases when the amount of solvent increases
(E) None of above

67. In which of the following equations, X represents an alpha particle?

(A) $^{238}U \rightarrow {}^{234}Th + X$
(B) $^1H^{35}Cl \rightarrow {}^{35}Cl^- + X$
(C) $^3H \rightarrow {}^3He + X$
(D) $^{11}Na \rightarrow {}^{11}Na^+ + X$
(E) $^{14}C \rightarrow {}^{14}N + X$

Questions 68–69: The following elements are listed in order of decreasing reactivity as they appear in the electrochemical series.

Ca, Na, Mg, Zn, Fe, H, Cu, Hg, Ag, Au

68. The element that is the best reducing reagent is

(A) H   (B) Zn   (C) Hg   (D) Na   (E) Mg

69. Of the following, the element that does not react with hydrochloric acid to produce hydrogen gas is

(A) Ca   (B) Au   (C) Fe   (D) Mg   (E) Zn

70. Which of the following expressions is correct for calculating the number of iron atoms in 5.0 grams of iron (Fe, mass number 55.9)?

(A) $5 \times 55.9 \times (6.02 \times 10^{23})$

(B) $\frac{10 \times 6.02 \times 10^{23}}{5 \times 55.9}$

(C) $\frac{5 \times (6.02 \times 10^{23})}{55.9}$

(D) $\frac{55.9}{5 \times (6.02 \times 10^{23})}$

(E) $\frac{5}{55.9 \times (6.02 \times 10^{23})}$

STOP!

If you finish before time is called, you may check your work on this section only. Do not turn to any other section in the test.

Practice Test 2 Answers

PART A and C							PART B	
#	Answer	#	Answer	#	Answer		#	Answer
1	B	25	B	49	A		101	True, True, No
2	E	26	B	50	A		102	True, False, No
3	A	27	C	51	C		103	False, True, No
4	D	28	D	52	C		104	False, True, No
5	B	29	D	53	D		105	True, False, No
6	D	30	D	54	E		106	True, True, Yes
7	E	31	B	55	B		107	True, False, No
8	C	32	D	56	D		108	True, True, Yes
9	A	33	D	57	B		109	True, False, No
10	C	34	E	58	B		110	True, True, No
11	E	35	D	59	A		111	False, False, No
12	B	36	E	60	D		112	True, True, Yes
13	D	37	B	61	A		113	False, True, No
14	A	38	B	62	B		114	True, True, Yes
15	A	39	E	63	C		115	True, True, Yes
16	E	40	C	64	D		116	
17	A	41	D	65	D			
18	A	42	B	66	B			
19	B	43	C	67	A			
20	E	44	B	68	D			
21	B	45	C	69	B			
22	C	46	E	70	C			
23	C	47	D	71				
24	C	48	C	72				

Calculation of the raw score

The number of correct answers: _____ = No. of correct

The number of wrong answers: _____ = No. of wrong

Raw score = No. of correct – No. of wrong x ¼ = _____

Score Conversion Table

Raw Score	Scaled Score	Raw Score	Scaled Score	Raw Score	Scaled Score
80	800	49	600	18	420
79	800	48	590	17	410
78	790	47	590	16	410
77	780	46	580	15	400
76	770	45	580	14	390
75	770	44	570	13	390
74	760	43	560	12	380
73	760	42	560	11	370
72	750	41	550	10	360
71	740	40	550	9	360
70	740	39	540	8	350
69	730	38	540	7	350
68	730	37	530	6	340
67	720	36	520	5	340
66	710	35	520	4	330
65	700	34	510	3	330
64	700	33	500	2	320
63	690	32	500	1	320
62	680	31	490	0	310
61	680	30	490	−1	310
60	670	29	480	−2	300
59	660	28	480	−3	300
58	660	27	470	−4	290
57	650	26	470	−5	280
56	640	25	460	−6	280
55	640	24	450	−7	270
54	630	23	450	−8	270
53	620	22	440	−9	260
52	620	21	440	−10	260
51	610	20	430		
50	600	19	420		

Explanations: Practice Test 2

1. **(B)** The electron configuration of Ca (Z = 20) is $1s^2 2s^2 2p^6 3s^2 3p^6 4s^2$. After losing the two $4s$ electrons (valence electrons), it becomes Ca^{2+} ion with electron configuration $1s^2 2s^2 2p^6 3s^2 3p^6$.

*See **Figure** 1.2 for this and next 4 questions.*

2. **(E)** Transition metals have partially filled d orbitals (d orbitals can hold up to 10 electrons). Only (E) has partially filled d orbitals ($3d^6$).

3. **(A)** The electron configuration of $1s^2 2s^2 2p^6 3p^1$ is of an excited state atom. The ground state electron configuration for an atom with 11 electrons (Na) should be $1s^2 2s^2 2p^6 3s^1$, $3s$ orbitals have lower energy than $3p$ orbitals and are filled before $3p$ orbitals for ground state atom.

4. **(D)** Potassium (K, Z = 19) atom has 19 electrons, the ground state electron configuration of potassium is $1s^2 2s^2 2p^6 3s^2 3p^6 4s^1$.

5. **(B)** The electron configuration $1s^2 2s^2 2p^6 3s^2 3p^6$ meet octet rule and represents a ground state noble gas (Ar, Z = 18).

6. **(D)** R–COO–R' (or RCOOR') is general formula for **esters**. **Table** 2.1 lists general structures, formulae and functional groups of major organic compound classes.

Table 2.1 *General Structures of Common Organic Compound Classes*			
Class	**General Structure**	**General Formula**	**Functional Group**
Alcohol	R—O．H	R–OH or ROH	–OH
Ether	R—O—R'	R–O–R or ROR'	–O–
Aldehyde	R—C(=O)—H	R–CHO or RCHO	–CHO
Ketone	R—C(=O)—R'	R–O–R' or ROR'	–C=O
Carboxylic acid	R—C(=O)—OH	R–COOH or RCOOH	–COOH
Ester	R—C(=O)—OR'	R–COO–R' or RCOOR'	–COO–
Alkyl halide	R—X	R–X or RX	–X (X = F, Cl, Br, I)

7. **(E)** R–CO–R' (RCOR') is general formula for **ketones** (–CO– represents C=O functional group).

8. **(C)** R–COOH (RCOOH) is general formula for **carboxylic acids**.

9. **(A)** Br_2 and Hg are liquid at room temperature. (B) Cl_2 and F_2 are gas at room temperature. (C) NH_4^+ and H_3O^+ may refer to aqueous or gaseous phase ions. (D) Fe and Co are solid metals at room temperature. (E) Diamond and graphite are solid at room temperature.

10. **(C)** A **coordinate covalent bond** (also called a dative covalent bond) is a covalent bond in which both electrons come from the same atom. Examples of coordinate bonds are H_3O^+ and NH_4^+. In H_2O molecule,

two H atoms form two O–H bonds by sharing electron with O atom, each H atom contribute one electron. O atom has two lone electron pair, which can attract a proton, H^+, and form the third O–H bond in which the two electrons are all from O atom. Although this O–H bond is indistinguishable from the other two, this third O–H bond is called **coordinate bond** (**Table** 1.2).

11. **(E) Allotrope**s are substances of two or more different physical forms in which an element can exist. Graphite, charcoal, and diamond are all allotropes of carbon.

12. **(B)** Cl_2 and F_2 are in the same group (VII, halogens) and are all strong oxidizer.

13. **(D)** A **pH indicator** (represented as HIn) is a weak organic **acid** or base that **exist** in more than one structural **form** (tautomers), of which at least one **form** is colored. In acid, pH indicator exists as protonated form (HIn); as pH increase, its deprotonated form (In^-) increases, and eventually, most protonated form converts to deprotonated form.

14. **(A)** An **Arrhenius acid** is defined as a substance that dissociates in water to form hydrogen ions (H^+). In other words, an Arrhenius acid increases the concentration of H^+ ions in an aqueous solution, and pH < 7.

15. **(A)** A solution with $[OH^-] < 10^{-7}$ is acidic ($[H^+] > 10^{-7}$), and has pH < 7. So the answer is the same as Question 14.

16. **(E)** is a **phase diagram** which shows different states of a substance at different temperature/pressure conditions.

17. **(A)** At fixed volume, pressure of an ideal gas is proportional to the absolute temperature. This relationship is defined by ideal gas law, $PV = nRT$, where P, V and T are pressure, volume and absolute temperature respectively; n is molar number of the gas and R is ideal gas constant. Only (A) depicts such relationship.

18. **(A) Kinetic energy** is proportional to absolute temperature as define in $\overline{K}_e = \frac{3}{2}kT$, where $\overline{K}_e$ is average kinetic energy, k is constant and T is absolute temperature. Similar to Question 17, (A) describe such relationship.

19. **(B)** At fixed temperature, the volume of an ideal gas is inversely proportional to its pressure (Boyle's Law). (B) describes such relationship.

20. **(E) Nuclear fusion** is the process that small atoms merge to form larger atom and release large amount of energy. The hydrogen fusion reaction found in stars is an example.

21. **(B)** In **beta decay** (β decay), a neutron converts to proton and emits an electron. This results in increase of atomic number by 1, but no change in mass number.

22. **(C)** In **alpha decay** (α decay), the nucleus emits an alpha particle (composed of 2 neutrons and 2 protons), resulting in a decrease in the atomic number by 2 and mass number by 4. The decay from ^{238}U to ^{234}Th is an example of alpha decay.

23. **(C)** The balanced equation is: $2HMnO_4 + 5H_2SO_3 \rightarrow 2MnSO_4 + 3H_2SO_4 + 3H_2O$.

    This set of questions tests your ability in **balancing chemical equations**. See **Table** 2.2 for details.

Table 2.2 *Balancing Chemical Equation*

The basic principles in balancing a chemical reaction equation are the **law of mass conservation and charge conservation**. Number of all atoms of the same elements in the left and right sides of the equation must be the same, and total charge (including sign) must be the same on the two sides. There is no specific approach can be used for all equations. You need to develop your own approach for balancing various equations.

To balance the equation $HMnO_4 + H_2SO_3 \rightarrow MnSO_4 + H_2SO_4$, follow the steps below:

> **Step 1**: Since this is a reduction–oxidation reaction, write the oxidation number for atoms involved in reduction (Mn) and oxidation (S): H Mn O$_4$ + H$_2$ S O$_3$ → Mn S O$_4$ + H$_2$ S O$_4$
> $\qquad\qquad$ +7 $\qquad$ +4 $\qquad$ +2 +6 $\qquad$ +6
>
> **Step 2**: Figure out changes in oxidation number for both reducing and oxidizing reagents. In this reaction, change of oxidation number of Mn is –5 (from +7 to +2), and change of oxidation number of S is +2 (from +4 to +6).
>
> **Step 3**: Add coefficients for the compounds containing the reducing and oxidizing elements. Add 2 before MnSO$_4$ first and 5 before H$_2$SO$_3$. For the right side, it's a bit complicated. First, add 2 before MnSO4 since Mn only appears once on the right side. This takes 2 S atoms. To balance S, add 3 before H$_2$SO$_4$.
>
> $\qquad$ 2HMnO$_4$ + 5H$_2$SO$_3$ → 2MnSO$_4$ + 3H$_2$SO$_4$ + H$_2$O
>
> **Step 4**: Balance other elements. Check the equation to see if other elements (H and O) are balanced. There are 23 O atoms on the left side, and 21 O atoms on the right side. There are 12 H atoms on the left side and 8 H atoms on the right side. To balance both the O and H atom numbers, a coefficient of 3 for H$_2$O will bring the entire equation to be balanced.
>
> $\qquad$ 2HMnO$_4$ + 5H$_2$SO$_3$ → 2MnSO$_4$ + 3H$_2$SO$_4$+ 3H$_2$O

24. **(C)** The balanced equation is: Fe$_2$O$_3$ + 3CO → 2Fe + **3CO$_2$**.

   Here is an example of balancing redox reaction equation:

   Step 1: Again, this is a reduction–oxidation reaction. Mark the oxidation number for elements involved in reduction and oxidation reaction

   $\qquad$ Fe$_2$O$_3$ + C O → Fe + C O$_2$
   $\qquad$ +3 $\qquad$ +2 $\qquad$ 0 $\quad$ +4

   Step 2: Figure out changes in oxidation number for both reducing and oxidizing reagents. In this reaction, change of oxidation number of Fe is –3 (from +3 to 0), and change of oxidation number of C is +2 (from +2 to +4).

   Step 3: Add coefficients for the compounds containing the reducing and oxidizing elements. Add 2 before Fe (there is no need to add coefficient before Fe2O3 since there is already 2 Fe atoms in the molecule). Add 3 before CO and CO$_2$.

   $\qquad$ Fe$_2$O$_3$ + 3CO → 2Fe + 3CO$_2$

   Step 4: Balance other elements. The only other element is O, it's balanced already.

25. **(B)** The balanced equation is: C$_2$H$_4$ + 3O$_2$ → **2CO$_2$ + 2H$_2$O**.

26. **(B)** This is an acid/base titration question. To completely titrate NaOH, amount of H$^+$ added equals to the amount of OH$^-$, i.e. V_a x C_a = V_b x C_b.

   C_b = (V_a x C_a)/ V_b = (10 mL x 1.0 M)/50 mL = 0.2 M

27. **(C)** In propyne, H–C≡C–CH$_3$, the triple bond has one sigma (s) bond and two pi (π) bonds (**Table** 1.13). Second, the two carbon atoms in the triple bond have sp hybridization, and the H atom and the third C connected to these two C atoms are in a straight line. This is similar to ethyne (H–C≡C–H) those four atoms are in a straight line.

28. **(D)** This question is about the concept of spontaneity of reaction and how to use the relationship between changes in free energy (ΔG), enthalpy (ΔH) and entropy (ΔS) to evaluate spontaneity of reaction (**Table** 2.3).

Table 2.3 *Spontaneity of Reaction*	

First, the spontaneity of reaction is established by ΔG as follow:

Sign of ΔG	Reaction Spontaneity
–	spontaneous
+	nonspontaneous
0	equilibrium

Second, ΔG is related to ΔH and ΔS by equation: $\mathbf{\Delta G = \Delta H - T\Delta S}$

Third, sign of ΔG may be determined by signs of ΔH and ΔS as follow:

Sign of ΔH	Sign of ΔS	Sign of ΔG	Reaction Spontaneity
–	+	–	Spontaneous at all T
+	–	+	Nonspontaneous at all T
+	+	+ or –	Spontaneous at high T
–	–	+ or –	Spontaneous at low T

29. **(D)** Concentration of H^+ ion in solution of pH = 5 is 10^{-5} M. Concentration of OH^- is $10^{-14}/10^{-5} = 10^{-9}$ M. Alternatively, pOH + pH = 14, pOH = 14 – pH = 9. Therefore, $[OH^-] = 10^{-9}$ M.

30. **(D)** This is a single replacement reaction: $Zn(s) + HCl\,(aq) \rightarrow \mathbf{H_2}(g) + \mathbf{ZnCl_2}(aq)$.

31. **(B)** The **empirical formula** of a chemical compound is the **simplest positive integer ratio** of atoms present in a compound. In contrast, the molecular formula identifies the number of each type of atom in a molecule. For example, the empirical formula of hydrogen peroxide, H_2O_2, would simply be HO.

    (A) ~~C_2H_3~~ and $C_4H_{10} \rightarrow C_2H_5$
    (B) CH and C_6H_6, correct answer
    (C) ~~CH_2~~ and $C_3H_8 \rightarrow C_3H_8$
    (D) ~~CH~~ and $C_2H_4 \rightarrow CH_2$
    (E) ~~CH_3~~ and $CH_4 \rightarrow CH_4$

32. **(D)** Gold foil experiment of Rutherford is a famous experiment in which a beam of alpha particles bombard a thin foil of gold. Rutherford made some very important conclusions based his observation (**Table** 2.4). However, the sign of nucleus was not discovered in this experiment.

Table 2.4 *Gold foil experiment of Rutherford*
Main observations of this experiment

Main observations of this experiment

   1. Almost all the alpha particles pass through the foil (A)
   2. Some alpha particles were deflected at different angles (B)
   3. Very few of the alpha particles were reflected after hitting the gold foil

Based on these observations, Rutherford made the following conclusions:

   1. Since most of the alpha particles passed straight through the gold foil without any deflection, most of the space within the atoms is empty (C);
   2. Since some of the alpha particles (which are big in size) were deflected by large angles or bounced backwards, they must have approached some positively charged region responsible for the deflection. This positively charged region is now called the nucleus (E);

> 3. Since very few alpha particles were deflected, the volume occupied by the central region (nucleus) is very small;
> 4. Since alpha particles (relatively dense) were deflected by the central volume of charge, almost the complete mass of the atom must be within the central volume.

33. **(D)** The diagram below (**Figure** 2.1) illustrates relationship between potential energy, activation energy of forward (E^f and $E_a{}^f$) and reverse (E^r and $E_a{}^r$) reaction. Use the equation $E^f + E_a^f = E^r + E_a{}^r = E_a$, the activation of reverse reaction is $E_a{}^r = E_a - E^r = 52 \text{ kJ/mol} - 15 \text{ kJ/mol} = 37 \text{ kJ/mol}$.

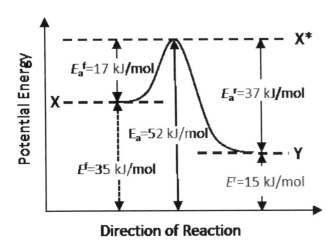

Figure 2.1 Potential energy, activation energy of forward and reverse reaction.

34. **(E)** This question tests your knowledge in precipitation reaction and solubility of salt. AgCl, PbBr$_2$, BaSO$_4$ are all insoluble. The following is the solubility rules:

Table 2.5 *Solubility Rules*
a. Salts containing Group I elements are soluble (Li$^+$, Na$^+$, K$^+$, Cs$^+$, Rb$^+$).
b. Salts containing the ammonium ion (NH$_4{}^+$) are soluble.
c. Salts containing nitrate ion (NO$_3{}^-$) are generally soluble.
d. Salts containing Cl$^-$, Br$^-$, I$^-$ are generally soluble. Important **exceptions** are halide salts of Ag$^+$, Pb^{2+}, and (Hg$_2$)$^{2+}$. Thus, **AgCl**, **PbBr$_2$**, and Hg$_2$Cl$_2$ are all insoluble.
e. Most silver salts are insoluble. AgNO$_3$ and Ag(C$_2$H$_3$O$_2$) are common soluble salts of silver.
f. Most sulfate salts are soluble. Important **exceptions** include **BaSO$_4$**, PbSO$_4$, Ag$_2$SO$_4$ and SrSO$_4$.
g. Most hydroxide salts are only slightly soluble. Hydroxide salts of Group I elements are soluble. Hydroxide salts of Group II elements (Ca, Sr, and Ba) are slightly soluble. Hydroxide salts of transition metals and Al^{3+} are insoluble. Thus, Fe(OH)$_3$, Al(OH)$_3$, Co(OH)$_2$ are insoluble.
h. Most sulfides of transition metals are highly insoluble. Thus, CdS, FeS, ZnS, Ag$_2$S are all insoluble.
i. Carbonates are frequently insoluble. Group II carbonates (Ca, Sr, and Ba) are insoluble. Some other insoluble carbonates include FeCO$_3$ and PbCO$_3$.
j. Chromates are frequently insoluble. Examples: PbCrO$_4$, BaCrO$_4$.
k. Phosphates are frequently insoluble. Examples: Ca$_3$(PO$_4$)$_2$, Ag$_3$PO$_4$.
l. Fluorides are frequently insoluble. Examples: BaF$_2$, MgF$_2$, PbF$_2$.

35. **(D)** When solid (and liquid) appears in an **aqueous equilibrium**, its concentration is considered to be 1; therefore, it is not included in the equilibrium constant expression.

36. **(E)** Catalyst can speed up reaction by **reducing the activation energy** of both reactant and products, so (E) is correct but (D) is incorrect. At constant temperature, catalyst does not change equilibrium constant, (A) is incorrect. Catalyst does not affect concentration of reactants or product, (B) and (C) are incorrect.

37. **(B)** Brønsted acid is defined as species which donates a proton and Brønsted base accepts proton. In the reaction $H_2O(l) + S^{2-} \rightleftharpoons HS^- + OH^-$, the species which give away a proton in the forward reaction is H_2O; and the species which give away a proton in the reverse reaction is HS^-.

38. **(B)** 1.0 M of NO_3^- solution is equivalent with 0.5 molar of $Pb(NO_3)_2$ solution, since there are two NO_3^- ions in the formula. 100 mL (0.1 L) of 0.5 M $Pb(NO_3)_2$ solution contains

    331 g/mol x (0.1 L x 0.5 mol/L) = 16.5 gram $Pb(NO_3)_2$.

39. **(E)** All natural radioactive decay process are spontaneous with a characteristic decay rate (also expressed as half–life). The decay product is an alpha (α), beta (β) particles or gamma (γ) rays

40. **(C)** 0.4 M NaCl solution yield highest concentration of ions in the solution, and cause the greatest boiling point elevation. This question is about **boiling point elevation**, which is one of **colligative properties.**

Table 2.6 *Boiling Point Elevation*

When a non–volatile solute is dissolved in water, the **boiling point elevation** of the resulted solution depends on the concentration of the solution:

$\Delta T_b = K_b \times m$ (1)

Where ΔT_b is the observed change in the boiling point (°C), K_b is the boiling point elevation constant (which is solvent–dependent), and m is the **molality**, or moles of solute per kilogram of solvent.

When the solute is a strong electrolyte, boiling point elevation is also dependent on the concentrations of all ions in the solution:

$\Delta T_b = i \times K_b \times m$ (2)

where i = van 't Hoff factor.

In this question, although $AlCl_3$ has highest i value, but the concentration (0.1 M) is lowest, and total concentration of all ions is 0.3 M. For 0.2 M $CaCl_2$ and 0.4 M **NaCl** solution, total concentration of all ions are 0.6 and **0.8** M respectively. For CH_3O, it does not hydrolyze in water ($i = 1$), and the concentration of solute is 0.5 M. Therefore, 0.4 M NaCl causes largest boiling point elevation.

Note: The concentration term in boiling point elevation equation is molality (m), not molarity (M). However, when comparing the concentration of solute in different solutions, molarity maybe used for comparison.

---

41. **(D)** When gaseous CO_2 convert to dry ice at low temperature, the randomness of system (entropy) decreases ($\Delta S < 0$). All other processes result in increase in randomness of system ($\Delta S > 0$).

42. **(B)** The balanced equation is: $3ClO^- \rightleftharpoons ClO_3^- + 2Cl^-$.

43. **(C)** No eating, drinking, chewing allowed any time in a chemistry laboratory. Choice A, B and E are common sense laboratory safety practice. For choice D, it's not required to pour liquid reagent over a sink, but it's a good practice if it can be done this way.

44. **(B)** The total pressure increases by six times due to addition of Ar, but partial pressure of H_2 does not change if the volume and temperature are not changes as defined by gas law, $P_i = n_i RT/V$.

45. **(C)** Follow the steps below to solve limiting reactant problem (also see **Table 1.6**):

    Step1: Balance the reaction equation $2Mg(s) + O_2(g) \rightarrow 2MgO(s)$

Step 2: Identify the limiting reactant. 48.6 grams of Mg equals to 2 mol of Mg, 64.0 grams of O_2 equals to 2 mol of O_2. Since the ratio of Mg to O_2 is 2:1 in the reaction, only 1 mol of O_2 is needed to oxidize 2 mol of Mg. Therefore, Mg is limiting reactant.

Step 3: Calculate amount of MgO produced. MgO produced is at 1:1 ration with Mg consumed. Therefore, 2 mol of MgO is produced, this equals to 80.6 grams of MgO.

46. **(E)** Use ideal gas law, $P_1V_1 = nRT_1$ (1) and $P_2V_2 = nRT_2$ (2).

Divide equation (1) by (2): $\dfrac{P_1V_1}{P_2V_2} = \dfrac{T_1}{T_2}$

Rearrange the equation: $V_2 = V_1 \times \dfrac{P_1}{P_2} \times \dfrac{T_2}{T_1}$

Plug in the numbers, the correct answer is E.

Note: temperature must be absolute temperature, and there is no need to covert volume to L.

47. **(D)** This question is about **Le Chatelier's principle**, also called **equilibrium law**.

Le Chatelier's principle states that if a chemical system at equilibrium experiences a change in concentration, temperature, volume, or partial pressure, then the equilibrium shifts to counteract the imposed change and a new equilibrium is established. Le Chatelier's principle can be used to predict the effect of a change in conditions on a chemical equilibrium (**Table** 1.5). Use the equilibrium in this problem as example: $A + B \rightleftharpoons AB + heat$

   I. When removing AB from the system, the system will shift to the direction at which more AB is produced to counteract the reduction of AB in the system. I is correct.

  II. When increase the temperature, to cancel the increase in temperature, the system will shift to the direction at which heat is absorbed. This is the reverse direction. II is incorrect.

 III. When adding more reactant A, the system will shift to the direction at which A is reduced. This is the forward direction. III is correct.

48. **(C)** This question test periodic property (electronegativity) of elements. Generally, in the same row (period), electronegativity increases from left to right; therefore, K has lower electronegativity than Ca and Fe. Second, within a group, electronegativity decrease downward. Use these two rules together, K has lower electronegativity than Si and Cl (**Figure** 2.2).

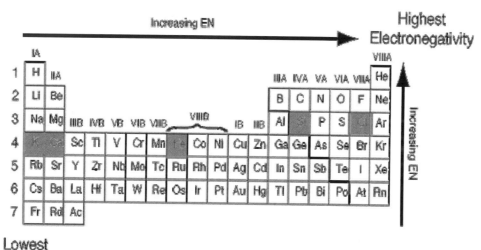

Figure 2.2 Periodic trends of electronegativity.

49. **(A)** Effusion is the process in which a gas escapes through a small hole. According to **Graham's law**, the rate of effusion of a gas is inversely proportional to the square root of the mass of its particles. Gases with a higher molecular mass effuse slower than gases with a lower molecular mass. Therefore, H_2 has the greatest rate of effusion since it has the smallest molar mass.

50. **(A)** is correct answer. **(B)** is incorrect since ideal gas particles are considered has no size (volume). **(C)** is incorrect since ideal gas particles travel in random manor. **(D)** is incorrect since ideal gas particles collide elastically. **(E)** is incorrect since volume of ideal gas change with both pressure and temperature. Only at standard pressure and temperature, 1 mole ideal gas occupies 22.4 liters.

51. **(C)** The periodic trend of size of atoms is shown below (**Figure** 2.3), decreasing from left to right in a row (period), and increasing down a column (group). Therefore, Cl has smallest size and Cs has largest size.

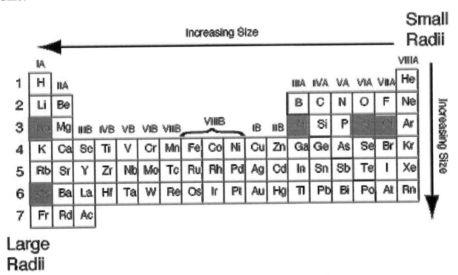

Figure 2.3 Periodic trend of atom size.

52. **(C)** This question is about **orbital hybridization, electron domain geometry** and **geometry of molecule** (**Table** 2.7). When central atom has sp^2 hybridization, there are two possible molecular geometry, trigonal planar (no lone pair) and bent (one lone pair). Therefore, **(C)** is correct answer.

Table 2.7 *Orbital Hybridization and Molecule Geometry*				
Number of electron groups	**Hybridization**	**Electron domain geometry**	**Number of lone pair**	**Molecular Geometry**
2	sp	linear	0	linear (BeF_2, CO_2)
3	sp^2	trigonal planar	0	trigonal planar (BF_3)
3	sp^2	trigonal planar	1	bent (SO_2)
4	sp^3	tetrahedral	0	tetrahedral (CH_4),
4	sp^3	tetrahedral	1	trigonal pyramid (NH_3),
4	sp^3	tetrahedral	2	bent (H_2O)

53. **(D)** Follow the steps below:

Step 1: Complete and balance the equation: $Cl_2(g) + 2KBr(aq) \rightarrow Br_2(g) + 2KCl(aq)$

Step 2: Determine amount of products. From the balanced equation, when 1 mol of Cl_2 completely reacts with excessive KBr, there will be 1 mol Br_2 and 2 mol KCl (Cl^-) produced.

54. **(E)** Aqueous ammonia is weak base. In basic solution, Al^{3+} tends to form $Al(OH)_3$ and precipitate since solubility of $Al(OH)_3$ is low.

55. **(B)** A neutral atom has the same number of electrons and protons. Therefore the atomic number of this element is 17 (Cl), and its full electron configuration is $1s^2 2s^2 2p^6 3s^2 3p^5$.

56. **(D)** Mass number (85) is sum of number of protons and neutrons; +1 charge indicates that the ion has 1 less electrons than proton. Only choice (D) meets these conditions. (A) has correct sum of neutrons and protons, but it has +2 charge; (B) has +1 charge, but the sum of neutrons and protons is 84; (C) has the correct mass number, but has no charge; and (E) has –1 charge and wrong mass number.

57. **(B)** The atomic number (number of protons) of Cl^- is 17 (**II is correct**). Now the mass number of the ion is 37, so the number of neutrons must be 20, therefore, I is incorrect. Neutral Cl should have 17 electrons, and Cl^- should have 18 electrons. Therefore, III is incorrect.

58. **(B)** The steps below are used to calculate number of hydrate water.

    Step 1: Mass of $CuCl_2$ = 134.5 g, and mass of hydrate water is 170.5 g – 134.5 g = 36.0 g.

    Step 2: Molar mass of $CuCl_2$ is 63.5 + 2 x 35.5 = 134.5 (g/mol), number of mol of $CuCl_2$ in the hydrate:

    $$134.5 \text{ g CuCl}_2 \times \frac{1 \text{ mol CuCl}_2}{134.5 \text{ g CuCl}_2} = 1.0 \text{ mol CuCl}_2.$$

    Number of mole of water is: $36 \text{ g H}_2O \times \frac{1 \text{ mol H}_2O}{18 \text{ g H}_2O} = 2.0 \text{ mol H}_2O$.

    Step 3: Ratio of mole between $CuCl_2$ to H_2O is 1:2, and X = 2.

    Step 4: Write hydrate formula: $CuCl_2 \cdot 2H_2O$.

59. **(A)** Refer to **Table** 1.11 for rules of oxidation number, oxidation number of S in the species are: SO_4^{2-} (+6), SO_3^{2-} (+4), SO_2 (+4), S_8 (0), H_2S (–2).

60. **(D) Molality**, also called **molal concentration**, is a measure of the concentration of a solute in a solution in terms of amount of substance in a **specified amount of mass** of the solvent. This contrasts with the definition of molarity which is based on a specified volume of solution.

61. **(A)** This problem may be solved in two ways

    Approach 1: add two reaction equations with enthalpy, and change reaction direction and sign of ΔH.

    $$\begin{array}{ll} 2NO_2 \rightarrow \cancel{N_2} + \cancel{2O_2} & \Delta H = -67.8 \text{ kilojoules} \\ + \quad \cancel{N_2} + \cancel{2O_2} \rightarrow N_2O_4 & \Delta H = +9.7 \text{ kilojoules} \\ \hline 2NO_2 \rightarrow N_2O_4 & \Delta H = -58.1 \text{ kilojoules} \end{array}$$

    Change direction and sign of ΔH of the combined reaction: $N_2O_4 \rightarrow 2NO_2 \quad \Delta H = +58.1$ kilojoules.

    Approach 2: change directions of both reactions and add two new reaction equations.

62. **(B)** Calculation of electrical potential from two half reaction is similar to calculation of enthalpy by combining two reaction, with a major difference. However, **when reaction coefficient is changes, the electrical potential does not change**.

    For this question, to obtain the combined reaction, follow the steps below:

    1) The first half reaction is reversed and the sign of potential is changed from + to –;

    $$Fe^{2+} \rightarrow Fe^{3+} + e^- \quad E° = -0.77 \text{ V}$$

    2) Multiply each term in the half reaction by 2, but the E remain unchanged;

    $$2Fe^{2+} \rightarrow 2Fe^{3+} + 2e^- \quad E° = -0.77 \text{ V}$$

3) Add the new half reaction with the second half reaction.

$$2Fe^{2+} \rightarrow 2Fe^{3+} + \cancel{2e^-} \qquad E° = -\mathbf{0.77}\ V$$
$$+\ Cl_2 + \cancel{2e^-} \rightarrow 2Cl^- \qquad E° = +1.36\ V$$
$$\overline{2Fe^{2+} + Cl_2 \rightarrow 2Fe^{3+} + 2Cl^- \quad E° = +0.59\ V}$$

63. **(C)** Glucose and methanol are not electrolyte, and their solutions are not electricity conductor. Acetic acid is weak acid, only partially hydrolyzed. HCl is strong acid, but the concentration is only 1/10 of $NaNO_3$. Therefore, 1.0 M $NaNO_3$ solution is the best electricity conductor.

64. **(D)** is correctly balanced.

(A) $Cl_2 + 2e^- \rightarrow Cl^-$ Not balanced $\rightarrow$ $Cl_2 + 2e^- \rightarrow \mathbf{2}Cl^-$
(B) $2e^- + Cu \rightarrow Cu^{2+}$ Incorrect $\rightarrow$ $Cu \rightarrow Cu^{2+} + 2e^-$
(C) $O_2 \rightarrow 2e^- + 2O^{2-}$ Incorrect $\rightarrow$ $O_2 + 4e^- \rightarrow 2O^{2-}$
(D) $Al^{3+} + 3e^- \rightarrow Al$ **Correctly balanced**
(E) $2H^+ + e^- \rightarrow H_2$ Not balanced $\rightarrow$ $2H^+ + \mathbf{2}e^- \rightarrow H_2$

65. **(D)** The molar mass of Pb atom and $PbSO_4$ are 207 and 303 respectively. Percent mass of Pb is 207/303 = 0.683 (68.3%).

66. **(B)** Solubility of gas increase with increases in pressure (B is correct), and decreases with increase in temperature (A is incorrect). Henry's Law describes relationship between solubility of gas with pressure at constant temperature (C is incorrect). Solubility is the amount of solute dissolves in **unit** volume of solvent, does not affected by amount of solvent, and (D) is incorrect.

67. **(A)** is correct answer. An alpha particle contains two neutrons and two protons. In an alpha decay, the mass number of daughter (decay product) is 4 less than the parent (original radioactive isotope).

(B) $^1H^{35}Cl \rightarrow {}^{35}Cl^- + X$ X is a hydrogen ion (H^+)
(C) $^3H \rightarrow {}^3He + X$ X is a beta particle (electron)
(D) $^{11}Na \rightarrow {}^{11}Na^+ + X$ X is an electron
(E) $^{14}C \rightarrow {}^{14}N + X$ X is a beta particle (electron)

68. **(D)** The elements in this question are listed in order of decreasing reactivity as in the electrochemical series. Since all elements listed are reducing reagents, their reactivity refers to their strength as reducing reagent. The strongest reducing reagent is the first element given, Ca, followed by Na and so on. Therefore, the best reducing reagent in the question is Na (Ca is not one of the choices).

69. **(B)** In the list, an element can react with the ion of the element to its right as in the electrochemical cell. An example is the reaction $Zn(s) + Cu^{2+}(aq) \rightarrow Zn^{2+}(aq) + Cu(s)$. Therefore, all elements to the left of H can react with acid to form hydrogen gas; all elements to the right of H will not react with HCl to form hydrogen gas.

70. **(C)** Conversion from mass to number of particle involves conversion from mass to mole first as follow:

Step 1: Number of mol of Fe in 5 grams of iron: 5.0/55.9 (mol)

Step 2: Number of mol of Fe x Avogadro number = (5.0/55.9 mol) x (6.02 x 10^{23} atoms/mol)
        = 5.0 x (6.02 x 10^{23})/55.9 (atoms).

101. **Correct answer: I True, II True, CE No**

Explanation: Both statements are true, but statement II does not explain Statement I. The fact that the bonds in H_2O is polar does not explain why the burning of H_2 gas is exothermic.

102. **Correct answer: I True, II False, CE No**

Explanation: The element with electron configuration of $[He]2s^2$ is Be (Z = 4), which is in the same row with fluorine. The radius of elements in the same row decreases from left to right, therefore, Be has larger radius than fluorine (F). Statement I is true. Be has +4 charge in its nucleus, while fluorine has +9 charges in its nucleus. The statement II is False.

103. **Correct answer: I False, II True, CE No**

Explanation: 1M $CaCl_2$ in solution causes greater boiling point elevation than 1 M NaCl because 1 M $CaCl_2$ yields higher total concentration of ions than NaCl (**Table** 2.4). Statement I is False and statement II is true.

104. **Correct answer: I False, II True, CE No**

Explanation: Henry's law states that solubility of gas in water increases with increase in pressure. Therefore, statement I is False. Statement II is true since vapor pressure of a liquid depends on temperature, but not surrounding air.

105. **Correct answer: I True, II False, CE No**

Explanation: At equilibrium, the rate of forward reaction equals to that of reverse reaction; that's why concentration of reactant and product does not change. Therefore, statement I is correct. However, both the forward and reverse reactions are still proceed, and the equilibrium reaction must be reversible. So, statement II is false.

106. **Correct answer: I True, II True, CE Yes**

Explanation: Greater surface area increases the contact between reactants, and increases rate of reaction. Therefore, both statement I and II are true, and statement II correctly explain statement I.

107. **Correct answer: I True, II False, CE No**

Explanation: HCl should not be collected with water displacement method since HCl is soluble in water (statement I is true). H–Cl bond is polar bond, and statement II is False.

108. **Correct answer: I True, II True, CE Yes**

Explanation: A liquid boils when its vapor pressure equals to the surrounding atmospheric pressure, and vapor pressure of a liquid is function of temperature. When a liquid is heated, its vapor pressure increases until it equals to surrounding air, and the liquid start to boil. The temperature at this point is boiling point. Therefore, boiling point of the liquid varies with surrounding pressure. When the surrounding atmospheric pressure is lowered, the boiling point is also lowered. The boiling point of water at 1 atm is 100°C. When pressure is lower than 1 atm (e.g. at high altitude), its boiling point is lower than 100°C.

109. **Correct answer: I True, II False, CE No**

Explanation: In an endothermic reaction, heat is added, and ΔH is positive (Statement I is true). This is because in endothermic reaction, the total potential energy of product is higher than that of reactant (**Figure** 2.4). The energy added is to increase the potential energy of product (Statement II is false).

110. **Correct answer: I True, II True, CE No**

Explanation: Volume of an ideal gas is inversely proportional to pressure (**Boyle's Law**), and proportional to temperature (in Kelvin, **Charles' Law**). Therefore, both Statement I and II are true, but they have no cause–effect relationship.

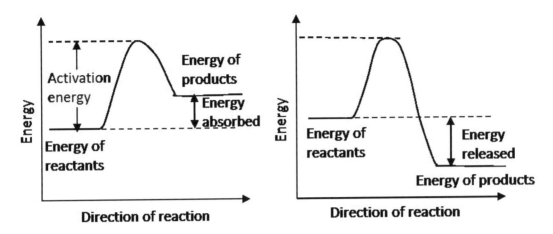

Figure 2.4 Potential energy of reactants and products in endothermic and exothermic reactions.

112. **Correct answer: I True, II True, CE No**

Explanation: Referring to the solubility rules in explanation for (**Table** 2.3), silver halogenates are insoluble in water (Statement II is true). This explains why the precipitation reaction of AgCl is not reversible (Statement II is true). Statement I explain statement II, but not the reverse.

113. **Correct answer: I False, II True, CE No**

Explanation: Catalyst increases rate of chemical reaction (Statement I is false) by lowering the activation energy of both reactant and product (Statement II is true).

114. **Correct answer: I True, II True, CE Yes**

Explanation: Electrolysis of water reaction is the reverse of the combustion reaction of hydrogen, which release heat (energy). You can use this fact to infer that electrolysis of water must need input of energy (Statement I is true). Then this is an endothermic reaction which absorbs energy because the potential energy of product is higher than that of reactant (Statement II is true and correctly explains Statement I).

115. **Correct answer: I True, II True, CE Yes**

Explanation: CH_3CH_2-OH and CH_3-O-CH_3 are different substances (ethanol and dimethyl ether) but have the same molecular formula (C_2H_6O, statement II is true). This fits the definition of isomers (statement I is true). And statement II is appropriate explanation of statement I.

SAT II Chemistry

Practice Test 3

Answer Sheet

Part A and C: Determine the correct answer. Blacken the oval of your choice completely with a No. 2 pencil.

1	Ⓐ Ⓑ Ⓒ Ⓓ Ⓔ	25	Ⓐ Ⓑ Ⓒ Ⓓ Ⓔ	49	Ⓐ Ⓑ Ⓒ Ⓓ Ⓔ
2	Ⓐ Ⓑ Ⓒ Ⓓ Ⓔ	26	Ⓐ Ⓑ Ⓒ Ⓓ Ⓔ	50	Ⓐ Ⓑ Ⓒ Ⓓ Ⓔ
3	Ⓐ Ⓑ Ⓒ Ⓓ Ⓔ	27	Ⓐ Ⓑ Ⓒ Ⓓ Ⓔ	51	Ⓐ Ⓑ Ⓒ Ⓓ Ⓔ
4	Ⓐ Ⓑ Ⓒ Ⓓ Ⓔ	28	Ⓐ Ⓑ Ⓒ Ⓓ Ⓔ	52	Ⓐ Ⓑ Ⓒ Ⓓ Ⓔ
5	Ⓐ Ⓑ Ⓒ Ⓓ Ⓔ	29	Ⓐ Ⓑ Ⓒ Ⓓ Ⓔ	53	Ⓐ Ⓑ Ⓒ Ⓓ Ⓔ
6	Ⓐ Ⓑ Ⓒ Ⓓ Ⓔ	30	Ⓐ Ⓑ Ⓒ Ⓓ Ⓔ	54	Ⓐ Ⓑ Ⓒ Ⓓ Ⓔ
7	Ⓐ Ⓑ Ⓒ Ⓓ Ⓔ	31	Ⓐ Ⓑ Ⓒ Ⓓ Ⓔ	55	Ⓐ Ⓑ Ⓒ Ⓓ Ⓔ
8	Ⓐ Ⓑ Ⓒ Ⓓ Ⓔ	32	Ⓐ Ⓑ Ⓒ Ⓓ Ⓔ	56	Ⓐ Ⓑ Ⓒ Ⓓ Ⓔ
9	Ⓐ Ⓑ Ⓒ Ⓓ Ⓔ	33	Ⓐ Ⓑ Ⓒ Ⓓ Ⓔ	57	Ⓐ Ⓑ Ⓒ Ⓓ Ⓔ
10	Ⓐ Ⓑ Ⓒ Ⓓ Ⓔ	34	Ⓐ Ⓑ Ⓒ Ⓓ Ⓔ	58	Ⓐ Ⓑ Ⓒ Ⓓ Ⓔ
11	Ⓐ Ⓑ Ⓒ Ⓓ Ⓔ	35	Ⓐ Ⓑ Ⓒ Ⓓ Ⓔ	59	Ⓐ Ⓑ Ⓒ Ⓓ Ⓔ
12	Ⓐ Ⓑ Ⓒ Ⓓ Ⓔ	36	Ⓐ Ⓑ Ⓒ Ⓓ Ⓔ	60	Ⓐ Ⓑ Ⓒ Ⓓ Ⓔ
13	Ⓐ Ⓑ Ⓒ Ⓓ Ⓔ	37	Ⓐ Ⓑ Ⓒ Ⓓ Ⓔ	61	Ⓐ Ⓑ Ⓒ Ⓓ Ⓔ
14	Ⓐ Ⓑ Ⓒ Ⓓ Ⓔ	38	Ⓐ Ⓑ Ⓒ Ⓓ Ⓔ	62	Ⓐ Ⓑ Ⓒ Ⓓ Ⓔ
15	Ⓐ Ⓑ Ⓒ Ⓓ Ⓔ	39	Ⓐ Ⓑ Ⓒ Ⓓ Ⓔ	63	Ⓐ Ⓑ Ⓒ Ⓓ Ⓔ
16	Ⓐ Ⓑ Ⓒ Ⓓ Ⓔ	40	Ⓐ Ⓑ Ⓒ Ⓓ Ⓔ	64	Ⓐ Ⓑ Ⓒ Ⓓ Ⓔ
17	Ⓐ Ⓑ Ⓒ Ⓓ Ⓔ	41	Ⓐ Ⓑ Ⓒ Ⓓ Ⓔ	65	Ⓐ Ⓑ Ⓒ Ⓓ Ⓔ
18	Ⓐ Ⓑ Ⓒ Ⓓ Ⓔ	42	Ⓐ Ⓑ Ⓒ Ⓓ Ⓔ	66	Ⓐ Ⓑ Ⓒ Ⓓ Ⓔ
19	Ⓐ Ⓑ Ⓒ Ⓓ Ⓔ	43	Ⓐ Ⓑ Ⓒ Ⓓ Ⓔ	67	Ⓐ Ⓑ Ⓒ Ⓓ Ⓔ
20	Ⓐ Ⓑ Ⓒ Ⓓ Ⓔ	44	Ⓐ Ⓑ Ⓒ Ⓓ Ⓔ	68	Ⓐ Ⓑ Ⓒ Ⓓ Ⓔ
21	Ⓐ Ⓑ Ⓒ Ⓓ Ⓔ	45	Ⓐ Ⓑ Ⓒ Ⓓ Ⓔ	69	Ⓐ Ⓑ Ⓒ Ⓓ Ⓔ
22	Ⓐ Ⓑ Ⓒ Ⓓ Ⓔ	46	Ⓐ Ⓑ Ⓒ Ⓓ Ⓔ	70	Ⓐ Ⓑ Ⓒ Ⓓ Ⓔ
23	Ⓐ Ⓑ Ⓒ Ⓓ Ⓔ	47	Ⓐ Ⓑ Ⓒ Ⓓ Ⓔ	71	Ⓐ Ⓑ Ⓒ Ⓓ Ⓔ
24	Ⓐ Ⓑ Ⓒ Ⓓ Ⓔ	48	Ⓐ Ⓑ Ⓒ Ⓓ Ⓔ	72	Ⓐ Ⓑ Ⓒ Ⓓ Ⓔ

Part B: On the actual Chemistry Test, the following type of question must be answered on a special section (labeled "Chemistry") at the lower left–hand corner of your answer sheet. These questions will be numbered beginning with 101 and must be answered according to the directions.

	I		II		CE
PART B					
	I		II		CE
101	T	F	T	F	◯
102	T	F	T	F	◯
103	T	F	T	F	◯
104	T	F	T	F	◯
105	T	F	T	F	◯
106	T	F	T	F	◯
107	T	F	T	F	◯
108	T	F	T	F	◯
109	T	F	T	F	◯
110	T	F	T	F	◯
111	T	F	T	F	◯
112	T	F	T	F	◯
113	T	F	T	F	◯
114	T	F	T	F	◯
115	T	F	T	F	◯
116	T	F	T	F	◯

Periodic Table of Elements

Material in this table may be useful in answering the questions in this examination

1 H 1.0079																	2 He 4.0026
3 Li 6.941	4 Be 9.012											5 B 10.811	6 C 12.011	7 N 14.007	8 O 16.00	9 F 19.00	10 Ne 20.179
11 Na 22.99	12 Mg 24.30											13 Al 26.98	14 Si 28.09	15 P 30.974	16 S 32.06	17 Cl 35.453	18 Ar 39.948
19 K 39.01	20 Ca 40.48	21 Sc 44.96	22 Ti 47.90	23 V 50.94	24 Cr 52.00	25 Mn 54.938	26 Fe 55.85	27 Co 58.93	28 Ni 58.69	29 Cu 63.55	30 Zn 65.39	31 Ga 69.72	32 Ge 72.59	33 As 74.92	34 Se 78.96	35 Br 79.90	36 Kr 83.80
37 Rb 85.47	38 Sr 87.62	39 Y 88.91	40 Zr 91.22	41 Nb 92.91	42 Mo 95.94	43 Tc (98)	44 Ru 101.1	45 Rh 102.91	46 Pd 106.42	47 Ag 107.87	48 Cd 112.41	49 In 114.82	50 Sn 118.71	51 Sb 121.75	52 Te 127.60	53 I 126.91	54 Xe 131.29
55 Cs 132.91	56 Ba 137.33	57 *La 138.91	72 Hf 178.49	73 Ta 180.95	74 W 183.85	75 Re 186.21	76 Os 190.2	77 Ir 192.2	78 Pt 195.08	79 Au 196.97	80 Hg 200.59	81 Tl 204.38	82 Pb 207.2	83 Bi 208.98	84 Po (209)	85 At (210)	86 Rn (222)
87 Fr (223)	88 Ra 226.02	89 Ac 227.03	104 Rf (261)	105 Db (262)	106 Sg (266)	107 Bh (264)	108 Hs (277)	109 Mt (268)	110 Ds (271)	111 Rg (272)	112 (277)						

*Lanthanide Series	58 Ce 140.12	59 Pr 140.91	60 Nd 144.24	61 Pm (145)	62 Sm 150.4	63 Eu 151.97	64 Gd 157.25	65 Tb 158.93	66 Dy 162.50	67 Ho 164.93	68 Er 167.26	69 Tm 168.93	70 Yb 173.04	71 Lu 174.97
Actinide Series	90 Th 232.04	91 Pa 231.04	92 U 238.03	93 Np 237.05	94 Pu (244)	95 Am (243)	96 Cm (247)	97 Bk (247)	98 Cf (251)	99 Es (252)	100 Fm (257)	101 Md (258)	102 No (259)	103 Lr (260)

Note: For all questions involving solutions, assume that the solvent is water unless otherwise stated.
Reminder: You may not use a calculator in this test!

Throughout the test the following symbols have the definitions specified unless otherwise noted.

H = enthalpy	atm = atmosphere(s)
M = molar	g = gram(s)
n = number of moles	J = joule(s)
P = pressure	kJ = kilojoule(s)
R = molar gas constant	L = liter(s)
S = entropy	mL = milliliter(s)
T = temperature	mm = millimeter(s)
V = volume	mol = mole(s)
	V = volt(s)

Chemistry Subject Practice Test 3

Part A

Directions for Classification Questions
Each set of lettered choices below refers to the numbered statements or questions immediately following it. Select the one lettered choice that best fits each statement or answers each question and then fill in the corresponding circle on the answer sheet. <u>A choice may be used once, more than once, or not at all in each set.</u>

Questions 1 – 4 refer to the following

    (A) Anion
    (B) Cation
    (C) Oxidizing reagent
    (D) Reducing reagent
    (E) Isotopes

1. Which is attracted to the anode in electrolysis.

2. Belong to the same element, but have different mass number.

3. The role that Cl_2 plays in the reaction:

    $H_2 + Cl_2 \rightarrow 2HCl$

4. Which migrates through the salt bridge to the anode electrode in a voltaic cell.

Questions 5 – 8 refer to the following laboratory devices

    (A) Calorimeter
    (B) Geiger counter
    (C) pH meter
    (D) Manometer
    (E) Voltmeter

5. Equipment used to measure acidity of solution

6. Equipment needed to determine the heat of reaction

7. Equipment needed to detect a nuclear leakage

8. Equipment needed to measure vapor pressure of a liquid

Questions 9 – 12 refer to the following substance

    (A) Strong acid
    (B) Strong base
    (C) Weak acid
    (D) Amphoteric substance
    (E) Buffer

9. Which can react as an acid or a base.

10. Which react with Zn swiftly to produce hydrogen gas.

11. A solution which has smaller change in pH compared to pure water when same amount acid or base is added.

12. Which is formed when alkali metal reacts with water.

Questions 13 – 16 refer to the following quantities

    (A) Atomic number
    (B) Mass number
    (C) Number of valence electrons
    (D) Number of neutrons
    (E) Maximum principal quantum number

13. Elements in the same row on periodic table has the same ___

14. Isotopes of an element has the same ___

15. Elements in the same group has the same ___

16. ^{14}C and ^{14}N has the same ___

GO ON TO THE NEXT PAGE

Questions 17 – 19 refer to the following statements

    (A) ΔH is positive
    (B) ΔH is negative
    (C) ΔS is positive
    (D) ΔS is negative
    (E) ΔG is positive

17. Which indicates a decrease in randomness

18. Which indicates a reaction is nonspontaneous

19. Which indicates an exothermic reaction

Questions 20 – 22 refer to the following groups of elements

    (A) Alkali metals
    (B) Alkaline earth metals
    (C) Transition metals
    (D) Halogens
    (E) Noble gases

20. Many has colored compounds due to partially filled d orbitals.

21. Are the most reactive nonmetals in a period.

22. Become octet state after losing two electrons.

Questions 23 – 25 refer to the following numbers

    (A) +5
    (B) +4
    (C) +1
    (D) –1
    (E) –4

23. Oxidation number of O in H_2O_2

24. Oxidation number of Cl in $NaClO_3$

25. Oxidation number of C in CH_4

GO ON TO THE NEXT PAGE

Practice Test 3———*Continued*

PLEASE GO TO THE SPECIAL SECTION AT THE LOWER LEFT–HAND CORNER OF PAGE 2 OF YOUR ANSWER SHEET LABELED CHEMISTRY AND ANSWER QUESTIONS 101–115 ACCORDING TO THE FOLLOWING DIRECTIONS.

<u>Part B</u>

Directions for Relationship Analysis Questions
Each question below consists of two statements, I in the left–hand column and II in the right–hand column. For each question, determine whether statement I is true or false and whether statement II is true or false and fill in the corresponding T or F circles on your answer sheet. *<u>Fill in circle CE only if statement II is a correct explanation of the true statement I.</u>

EXAMPLES:

I		II
EX1. The nucleus in an atom has a positive charge.	BECAUSE	Proton has positive charge, neutron has no charge.

SAMPLE ANSWERS

	I	II	CE
EX1	● Ⓕ	● Ⓕ	●

101. An element whose atoms have an outer electron configuration $3s^1$ shows metallic properties — BECAUSE — metallic elements readily gain electron.

102. An endothermic reaction has a positive ΔH value — BECAUSE — in an endothermic reaction, the total enthalpy (heat content) of the product is greater than that of the reactants.

103. A buret is normally used in volumetric titrations — BECAUSE — a buret can accurately measure the volumes of a solution delivered.

104. An electrolytic cell makes a nonspontaneous redox reaction occur — BECAUSE — An electrolytic cell uses an external electrical energy to drive a redox reaction.

105. The halogens all form stable diatomic molecules — BECAUSE — they all need one electron to fill their outermost shell.

106. 3 kilograms equal to 3000 grams — BECAUSE — the prefix kilo– means "one thousandth".

GO ON TO THE NEXT PAGE

85

107. An increase in temperature will cause a gas to expand   BECAUSE   when temperature increases, the kinetic energy of the molecules in the gas increases.

108. A catalyst makes a reaction faster   BECAUSE   a catalyst lessens the enthalpy change of the reaction.

109. The ideal gas law does not hold under low temperature and high pressure   BECAUSE   interaction between particles cannot be neglected under low temperature and high pressure.

110. Nitrogen gas will have a greater rate of effusion than oxygen gas   BECAUSE   nitrogen diatomic molecule has triple covalent bond, while oxygen diatomic molecule has double covalent bond.

111. Propane is compound   BECAUSE   the covalent bond between carbon and hydrogen in propane cannot be broken chemically.

112. A mixture of two different liquids can be separated via distillation   BECAUSE   different liquids boil at different temperature.

113. Isotopes have different atomic numbers   BECAUSE   isotopes have different numbers of neutrons.

114. When salt is added to water, the melting point becomes lower   BECAUSE   the salt dissolved in water lower the vapor pressure of water.

115. Radiation and radioisotopes can have beneficial uses   BECAUSE   radioisotopes and radiation can be used for radio dating, radiotracers, and food preservation.

GO ON TO THE NEXT PAGE

Part C

Directions for Five–Choice Completion Questions
Each of the questions or incomplete statements below is followed by five suggested answers or completions. Select the one that is best in each case and then fill in the corresponding circle on the answer sheet.

26. Given the bond energy below:

    Cl–Cl 243 kJ/mol
    H–H 436 kJ/mol
    H–Cl 431 kJ/mol

    Which of the following is the enthalpy change of the reaction?

    $H_2(g) + Cl_2(g) \rightarrow 2HCl(g)$

    (A) 0 kJ/mol
    (B) +183 kJ/mol
    (C) –183 kJ/mol
    (D) –1110 kJ/mol
    (E) +1110 kJ/mol

27. A certain amount of carbon required 13 grams of oxygen to be converted into carbon monoxide, CO. If the same amount of carbon were to be converted into carbon dioxide, CO_2, the amount of oxygen required would be

    (A) 6.5 grams
    (B) 13 grams
    (C) 26 grams
    (D) 52 grams
    (E) Indeterminable from given information

28. Which of the following is NOT part of the Atomic Theory?

    (A) Compounds form when atoms combine in whole number ratios
    (B) All atoms of the same element are alike
    (C) All matter is composed of atoms
    (D) Atoms cannot be created, subdivide, destroyed chemically.
    (E) An atom can be converted to another atom.

29. Which of the following correctly describes the geometry of the molecule?

    (A) CO_2 has a bent geometry
    (B) H_2O has a linear geometry
    (C) CH_4 has a tetrahedral geometry
    (D) BF_3 has a pyramidal geometry
    (E) NH_3 has a trigonal planar geometry

30. One atom of element X has an atomic number of 9 and a mass number of 20; one atom of element Y has an atomic number of 10 and a mass number of 20. Which of the following statements about these two atoms is true?

    (A) They are isotopes.
    (B) They are isomers.
    (C) They are isoelectronic.
    (D) They contain the same number of neutrons in their nucleus.
    (E) They contain the same total number of protons plus neutrons in their nucleus.

31. Each of the following system is at equilibrium in a closed container. A decrease in the total volume of each container increases the number of moles of product(s) for which system?

    (A) $Fe_3O_4(s) + 4H_2(g) \rightleftharpoons 3Fe(s) + 4H_2O(g)$
    (B) $H_2(g) + Cl_2(g) \rightleftharpoons 2HCl(g)$
    (C) $CO(g) + H_2O(g) \rightleftharpoons CO_2(g) + H_2(g)$
    (D) $2NO(g) + O_2(g) \rightleftharpoons 2NO(g)$
    (E) $2NH_3(g) \rightleftharpoons N_2(g) + 3N_2(g)$

GO ON TO THE NEXT PAGE

32. A 250 milliliter solution of $AgNO_3$ was treated with excess NaCl. The total amount of AgCl precipitated is 14.3 grams. What was the molarity of Ag^+ in the original solution?

    (A) 0.05 M
    (B) 0.10 M
    (C) 0.20 M
    (D) 0.40 M
    (E) 0.80 M

33. Which of the following processes will produce an aqueous solution with a molarity of 1.0 M for the solute?

    (A) 73 grams of HCl (molar mass = 36.5) dissolved to make 2.5 liters of solution
    (B) 360 grams of $C_6H_{12}O_6$ (molar mass = 180) dissolved to make 2.0 liters of solution
    (C) 94 grams of K_2O (formula mass = 94) dissolved to make 1.0 liters of solution
    (D) 24 grams of LiOH (formula mass = 24) dissolved to make 1.25 liters of solution
    (E) 40 grams of HF (molar mass = 20) dissolved to make 2.50 liters of solution

34. There is no insoluble precipitate formed in which of the following double replacement reactions?

    (A) $BaCl_2(aq) + Na_2SO_4(aq) \rightarrow$
    (B) $HgCl_2(aq) + Na_2S(aq) \rightarrow$
    (C) $Pb(NO_3)_2 (aq) + CuSO_4(aq) \rightarrow$
    (D) $AgNO_3(aq) + KI(aq) \rightarrow$
    (E) $HBr(aq) + Ca(OH)_2(aq) \rightarrow$

35. Which of the following aqueous solutions has the greatest electricity conductivity?

    (A) 1.0 M acetic acid
    (B) 1.0 M hydrochloric acid
    (C) 1.0 M hydrofluoric acid
    (D) 1.0 M sucrose
    (E) 1.0 M methanol

36. Which of the following statements explains why many of the complexes of the transition elements have color?

    (A) They have high formation constants.
    (B) They absorb light of some visible wavelengths.
    (C) They absorb light of some ultraviolet wavelengths.
    (D) They have unpaired electrons.
    (E) They are soluble in water.

37. Which of the following increase the rate at which a solid dissolves in water?

    I. Grounding the solid to powder before adding into the water
    II. Heating the mixture
    III. Stirring the mixture

    (A) I only
    (B) II only
    (C) I and II only
    (D) II and III only
    (E) I, II, and III

38. Which of the following quantities does not equal to "one mole"?

    (A) 22.4 L of helium gas at STP
    (B) 6.02×10^{23} silica atoms
    (C) 44 grams of carbon dioxide gas
    (D) 32 grams of water vapor
    (E) 40 grams of metal calcium

39. Which of the following statements regarding empirical formulas is FALSE?

    (A) The empirical formula for butyne is C_2H_3
    (B) The empirical formula for ammonia is NH_3
    (C) The empirical formula of CH_2O is $C_6H_{12}O_6$
    (D) Ionic compounds are written as empirical formulas
    (E) The empirical and molecular formulas for methane are the same

GO ON TO THE NEXT PAGE

40. Which of the following molecular formula represents a compound which contains 26% nitrogen and 74% oxygen by mass?

(A) NO
(B) NO_2
(C) N_2O
(D) N_2O_3
(E) N_2O_5

41. How many grams of Fe_2O_3 (formula mass = 160) can be formed from the rusting of 223 grams of Fe with excess oxygen according to the reaction:

$4Fe + 3O_2 \rightarrow 2Fe_2O_3$

(A) 271 grams
(B) 320 grams
(C) 446 grams
(D) 480 grams
(E) 640 ggram

42. Sodium and chlorine react according to the following reaction: $2Na + Cl_2 \rightarrow 2NaCl$. If the reaction starts with 5.0 moles of Na and 3.0 moles of Cl_2, which statement below is true?

(A) Cl_2 is the excess reagent and 3.0 moles of NaCl will be produced
(B) Na is the excess reagent and 6.0 moles of NaCl will be produced
(C) There will be 2.0 moles of Na left after the reaction is completed
(D) Na is the limiting reagent and 5.0 moles NaCl will be produced
(E) All Na and Cl2 will be consumed after the reaction is completed, and 5 moles of NaCl will be produced

43. Which of the following is a correct Lewis dot structure of S^- ion?

(A) $\left[:\overset{\bullet\bullet}{\underset{\bullet\bullet}{S}}: \right]^-$

(B) $\left[:\overset{\bullet}{\underset{\bullet\bullet}{S}}: \right]^-$

(C) $\left[:\overset{\bullet}{\underset{\bullet}{S}}: \right]^-$

(D) $\left[:\overset{\bullet}{\underset{\bullet}{S}}\bullet \right]^-$

(E) None of above

44. Given the reaction:

$3H_2(g) + N_2(g) \rightleftharpoons 2NH_3(g) +$ heat energy.

Which of the following would drive the equilibrium to the reverse direction of ammonia formation?

(A) Removing ammonia from the reaction
(B) Increasing the temperature of the system
(C) Increasing the pressure on the system
(D) Adding nitrogen gas
(E) Adding hydrogen gas

45. When HS^- acts as a Brønsted acid, which of the following is formed?

(A) S^{2-}
(B) H^+
(C) H_2S
(D) H_2S_2
(E) H_3S^+

GO ON TO THE NEXT PAGE

46. $Al_2(C_2O_4)_3(s) \rightarrow Al_2O_3(g) + CO(g) + CO_2(g)$

According to the equation for the reaction represented above, what is the mole ratio of CO to CO_2 that is produced by the decomposition of aluminum oxalate?

(A) 1:1
(B) 1:2
(C) 1:3
(D) 2:1
(E) 3:1

47. If the pressure on an ideal gas is doubled, the volume of the gas will be

(A) Doubled
(B) The same
(C) Halved
(D) Quartered
(E) Quadrupled

48. $Ca_3(PO_4)_2(s) \rightleftharpoons 3Ca^{2+}(aq) + 2PO_4^{3-}(aq)$

Which of the following is the correct solubility product of the dissolution equilibrium of calcium phosphate?

(A) $K_{sp} = [Ca^{2+}][PO_4^{3-}]$

(B) $K_{sp} = [Ca^{2+}]^2[PO_4^{3-}]^3$

(C) $K_{sp} = [Ca^{2+}]^3[PO_4^{3-}]^2$

(D) $K_{sp} = [Ca^{2+}][PO_4^{3-}] / [Ca_3(PO_4)_2]$

(E) $K_{sp} = [Ca^{2+}]^3[PO_4^{3-}]^2 / [Ca_3(PO_4)_2]$

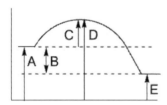

49. In the diagram shown above, which letter represents the enthalpy change of the reaction (ΔH)?

(A) A (B) B (C) C (D) D (E) E

50. The conversion of $Cr_2O_7^{2-}$ to Cr^{3+} during a chemical reaction is an example of

(A) hydrolysis
(B) displacement
(C) reduction
(D) oxidation
(E) decomposition

51. Molten KBr is allowed to undergo the process of electrolysis. Which of the following represents the reaction occurring at the anode?

(A) $K^+ + e^- \rightarrow K(s)$
(B) $K(s) \rightarrow K^+ + e^-$
(C) $2Br^- \rightarrow Br_2 + 2e^-$
(D) $Br_2 \rightarrow 2Br^- + 2e^-$
(E) $K^+ \rightarrow K(s) + e^-$

52. Which of the following will happen when sodium sulfate is added to a saturated solution of $PbSO_4$ that is at equilibrium?

$PbSO_4(s) \rightleftharpoons Pb^{2+}(aq) + SO_4^{2-}(aq)$

(A) The concentration of lead ions will increase
(B) The reaction will shift to the right
(C) The K_{sp} value increases
(D) The K_{sp} value decreases
(E) The K_{sp} value remains unchanged

53. To prepare 0.5 molar of sodium hydroxide solution, which of the following procedures is correct?

(A) Dissolve 20 g NaOH in 0.5 kg of water
(B) Dissolve 20 g NaOH in 1.0 kg of water
(C) Dissolve 20 g NaOH in water, and dilute to a final volume of 0.5 liter
(D) Dissolve 20 g NaOH in water, and dilute to a final volume of 1.0 liter
(E) Dissolve 20 g NaOH in water, and dilute to a final volume of 2.0 liter

GO ON TO THE NEXT PAGE

54. Which pair of the following compounds have polar bonds but are both nonpolar molecules?

    (A) H_2 and CH_4
    (B) H_2O and NH_3
    (C) SO_2 and CO_2
    (D) NaCl and AgI
    (E) CH_4 and CO_2

55. How many milliliters of 1.5 M hydrochloric acid is needed to titrate 45.0 milliliters of 1.0 M NaOH solution?

    (A) 15.0 mL
    (B) 30.0 mL
    (C) 45.0 mL
    (D) 35.0 mL
    (E) 20.0 mL

56. Which of the following statements regarding isotopes of an element is INCORRECT?

    (A) Isotopes have the same number of protons
    (B) Isotopes have the same atomic number
    (C) Isotopes differ in mass number
    (D) Isotopes differ in number of neutrons present
    (E) Isotopes differ in their nuclear charge

57. The ground state electron configuration of the silicon atom is characterized by which of the following?

    I. Partially filled $3p$ orbitals
    II. The presence of unpaired electrons
    III. Six valence electrons

    (A) I only
    (B) II only
    (C) I and II only
    (D) I and III only
    (E) I, II and III

58. What is the concentration of hydroxide ion, OH^-, in a solution with a pH of 11 at 25°C?

    (A) 1×10^{-1} M
    (B) 1×10^{-3} M
    (C) 1×10^{-5} M
    (D) 1×10^{-7} M
    (E) 1×10^{-11} M

59. Which of the following statements is INCORRECT regarding a straight–chained organic compound with the molecular formula of C_3H_8?

    (A) It is a saturated hydrocarbon.
    (B) Its empirical formula is the same as the molecular formula.
    (C) It is a gas at STP.
    (D) It is a nonpolar molecule.
    (E) The three carbon atoms are aligned in a straight line.

60. Which of the following is the weakest attractive force?

    (A) van der Waals force
    (B) Coordinate covalence bond
    (C) Polar covalence bond
    (D) Nonpolar covalence bond
    (E) Ionic bond

61. Which of the following statements is INCORRECT?

    (A) Melting is an endothermic process
    (B) Deposition is the reverse process of sublimation
    (C) Heat is released in condensation
    (D) Heat is needed to rise temperature in evaporation
    (E) Freezing is a exothermic process

GO ON TO THE NEXT PAGE

62. A triple bond in CH≡CH may be described as:

(A) Two sigma bonds and one pi bond
(B) Two sigma bonds and two pi bonds
(C) One sigma bond and two pi bonds
(D) Three sigma bonds
(E) Three pi bond

63. From left to right across a row (period) on the periodic table, which of the following characteristics of elements decrease?

(A) First ionization energy.
(B) Number of charge in nucleus
(C) Electronegativity
(D) The ability to gain electrons
(E) Metallic character

64. $Cu(s) + NO_3^- (aq) + H^+ (aq) \rightarrow Cu^{2+}(aq) + NO_2(g) + H_2O(l)$

Which of the following is true regarding the reaction above?

(A) $Cu(s)$ is oxidized
(B) $H^+(aq)$ is oxidized
(C) $Cu(s)$ is reduced
(D) $NO_3^- (aq)$ is reduced
(E) $NO_3^- (aq)$ is oxidized

65. What is the correct formula for iron(III) sulfate?

(A) $FeSO_4$
(B) $Fe_2(SO_4)_3$
(C) $Fe(SO_4)_3$
(D) Fe_3SO_4
(E) $Fe_3(SO_4)_2$

66. To prepare 2 liter of 0.5 M NaOH solution, how many milliliters of 10.0 M stock solution is needed?

(A) 10 mL
(B) 50 ml
(C) 100 mL
(D) 200 mL
(E) 1000 mL

67. A radioactive substance decays from 100 grams to 6.25 grams in 100 days. What is the half–life of this radioactive substance?

(A) 6.25 days
(B) 12.5 days
(C) 25 days
(D) 50 days
(E) 100 days

68. Which of the following statements about bonding is correct?

(A) Only van der Waals forces exist between polar molecules
(B) Dipoles are the result of the equal sharing of electrons
(C) $Cu(s)$ is a network covalent solid
(D) Hydrogen bonds exist between the molecules of HCl
(E) $NaCl(aq)$ has attraction between the molecules and the ions

GO ON TO THE NEXT PAGE

69. Which of the following choices demonstrate amphoterism?

    I. $HCl + H_2O \rightarrow H_3O^+ + Cl^-$ and $H_2O + NH_3 \rightarrow OH^- + NH_4^+$
    II. $HS^- + HCl \rightarrow Cl^- + H_2S$ and $HS^- + NH_3 \rightarrow NH_4^+ + S^{2-}$
    III. $HCl + NaOH \rightarrow NaCl + H_2O$ and $NaCl + H_2O \rightarrow HCl + NaOH$

    (A) I only
    (B) II only
    (C) III only
    (D) I and II only
    (E) I and III only

70. Which statement below is INCORRECT regarding balanced equations?

    (A) $C + O_2 \rightarrow CO_2$ is balanced and is a synthesis reaction
    (B) $CaCO_3 \rightarrow CaO + CO_2$ is balanced and is a decomposition reaction
    (C) $Na + Cl_2 \rightarrow NaCl$ is not balanced but demonstrates a synthesis reaction
    (D) $KI + Pb(NO_3)_2 \rightarrow PbI_2 + KNO_3$ is balanced and is a single replacement reaction
    (E) $2H_2O \rightarrow 2H_2 + O_2$ is balanced and demonstrates a redox reaction

STOP!

If you finish before time is called, you may check your work on this section only. Do not turn to any other section in the test.

Practice Test 3 Answers

\#	Answer	\#	Answer	\#	Answer
PART A and C					
1	A	25	E	49	B
2	E	26	C	50	C
3	C	27	C	51	C
4	A	28	E	52	E
5	C	29	C	53	D
6	A	30	E	54	E
7	B	31	D	55	B
8	D	32	D	56	E
9	D	33	B	57	C
10	A	34	E	58	B
11	E	35	B	59	E
12	B	36	B	60	A
13	E	37	E	61	D
14	A	38	D	62	C
15	C	39	C	63	E
16	B	40	B	64	A
17	D	41	B	65	B
18	E	42	D	66	C
19	B	43	B	67	C
20	C	44	B	68	E
21	D	45	A	69	D
22	B	46	A	70	D
23	D	47	C	71	
24	A	48	C	72	

PART B	
\#	Answer
101	True, False, No
102	True, True, Yes
103	True, True, Yes
104	True, True, Yes
105	True, True, Yes
106	True, False, No
107	True, True, Yes
108	True, False, No
109	True, True, Yes
110	True, True, No
111	True, False, No
112	True, True, Yes
113	False, True, No
114	True, True, Yes
115	True, True, Yes
116	

Calculation of the raw score

The number of correct answers: _____ = No. of correct

The number of wrong answers: _____ = No. of wrong

Raw score = No. of correct – No. of wrong x ¼ = _____

Score Conversion Table

Raw Score	Scaled Score	Raw Score	Scaled Score	Raw Score	Scaled Score
80	800	49	600	18	420
79	800	48	590	17	410
78	790	47	590	16	410
77	780	46	580	15	400
76	770	45	580	14	390
75	770	44	570	13	390
74	760	43	560	12	380
73	760	42	560	11	370
72	750	41	550	10	360
71	740	40	550	9	360
70	740	39	540	8	350
69	730	38	540	7	350
68	730	37	530	6	340
67	720	36	520	5	340
66	710	35	520	4	330
65	700	34	510	3	330
64	700	33	500	2	320
63	690	32	500	1	320
62	680	31	490	0	310
61	680	30	490	−1	310
60	670	29	480	−2	300
59	660	28	480	−3	300
58	660	27	470	−4	290
57	650	26	470	−5	280
56	640	25	460	−6	280
55	640	24	450	−7	270
54	630	23	450	−8	270
53	620	22	440	−9	260
52	620	21	440	−10	260
51	610	20	430		
50	600	19	420		

Explanations: Practice Test 3

1. **(A)** In electrolysis cell, anions are attracted to anode which is positively charged because electrons are withdrawn by a battery (**Table** 1.1).

2. **(E)** Isotopes of an element have the same number of proton (atomic number), but different number of neutrons, and subsequently, different mass number.

3. **(C)** The reaction $H_2 + Cl_2 \rightarrow 2HCl$ is reduction–oxidation reaction, H_2 is oxidized by Cl_2. Therefore, H_2 is reducing reagent, and Cl_2 is oxidizing reagent.

4. **(A)** In voltaic cell, anode loses mass and produce cation, anions travel to anode to balance the positive charge produced (**Table** 1.1).

5. **(C)** Acidity of a solution is expressed as pH, which is measured with **pH meter**.

6. **(A) Calorimeter** is device used to measure heat in a chemical or physical process.

7. **(B)** A **Geiger counter** detects radioactivity.

8. **(D)** A **gas manometer** can be used to measure vapor pressure of a liquid (**Figure** 3.1).

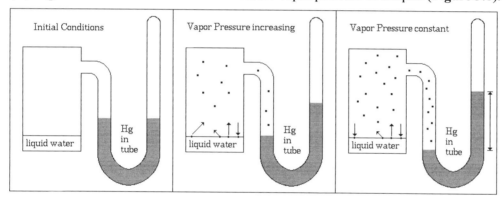

Figure 3.1 Use a manometer to measure vapor pressure of water.

9. **(D)** An **amphoteric substance**, by definition, is one which can act as either an acid or a base.

10. **(A)** A strong acid such as HCl or H_2SO_4 can react swiftly with Zn to produce H_2 gas. This is a single displacement reaction such as:

    $Zn(s) + HCl(aq) \rightarrow ZnCl_2(aq) + H_2(g)$

11. **(E)** A **buffer** is made by a weak acid or weak base together with its conjugate. When hydrogen ions are added to a buffer, they will be neutralized by the base in the buffer. Hydroxide ions will be neutralized by the acid. The neutralization reactions cancels the impact of addition of acid, and the pH of solution does not change significantly.

12. **(B)** All **alkali metals** (such as Na and K) react with water to produce hydrogen gas and strong base in solution as follow:

    $M + H_2O \rightarrow H_2 + MOH$

13. **(E)** All elements in a row have same number of layers of electron shells, or **maximum principal quantum number n**.

14. **(A)** Isotopes of an element have the same number of protons (**atomic number**).

15. **(C)** Chemical properties of elements in the same group are similar because they have same **number of valence electrons**.

16. **(B)** ^{14}C and ^{14}N belong to different elements, and have different number of protons in their nucleus. But they have same **mass number** (14), which is the sum of numbers of protons and neutrons.

17. **(D)** Decrease in entropy, $-\Delta S$, indicates decrease in randomness.

18. **(E)** Increase in free energy, $+\Delta G$, indicates the reaction (process) is nonspontaneous (**Table** 2.3).

19. **(B)** Negative change in enthalpy, $-\Delta H$, indicates the reaction release heat, and it's exothermic reaction.

20. **(C)** Many transition metals have color when forming compounds, this is caused by partially filled d orbitals.

21. **(D)** Halogens are the most reactive nonmetals in a row (period).

22. **(B)** Alkaline earth metals (IIA group) has two valence electrons, and tend to lose them to obtain octet electron configurations.

23. **(D)** To correctly assign **oxidation number** (O.N.) of an atom in a compound, a series of rules must be followed (**Table** 1.11). O.N. of O usually is –2 with a few exceptions, H_2O_2 is one exception in which O.N. of O is –1.

24. **(A)** Chlorine has multiple O.N.s, –1, 0, +1, +3, +5, +7. Chlorine can form several types of polyatomic anions with different O.N.s, such as in ClO^- (+1), ClO_2^- (+3), ClO_3^- **(+5)** and ClO_4^- (+7).

25. **(E)** Carbon in methane (CH_4) has O.N. –4. Carbon and nitrogen both have multiple O.N.s. It's commonly done by calculating O.N. of C or N from other elements in the compounds or ions.

    For example, in ethane (C_2H_6), O.N. of H is +1, and there are 6 H atoms and 2 C atoms. To make the total O.N. to be zero, O.N. of C is –3.

Table 3.1 *Oxidation numbers of carbon and nitrogen*

Substance	Formula	Oxidation State of (c)	Substance	Formula	Oxidation State of Nitrogen
Methan	CH_4	-4	Ammonia	NH_3	-3
Ethane	C_2H_6	-3	Hydrazine	N_2H_4	-2
Ethene	C_2H_4	-2	Hydride	N_2H_2	-1
Ethyne	C_2H_2	-1	Dinitrogen gas	N_2	0
Dichloromethane	CH_2Cl_2	0	Nitrous oxide	N_2O	+1
Chloroform	$CHCl_3$	+2	Nitric oxide	NO	+2
Oxalic acid	$(COOH)_2$	+3	Dinitrogen trioxide	N_2O_3	+3
Carbon TetraChloride	CCl_4	+4	Nitrogen dioxide	NO_2	+4
Carbon dioxide	CO_2	+4	Nitrogen Pentoxide	N_2O_5	+5

26. **(C)** Bond energy is the energy needed to break a bond. Heat of a reaction can be calculated from bond energy of all reactants and products. As in the reaction: $H_2(g) + Cl_2(g) \rightarrow 2HCl(g)$

    $\Delta H_{reaction} = \Sigma \Delta H_{bond\ broken} + \Sigma \Delta H_{bond\ formed}$
    $= (234\ kJ/mol + 436\ kJ/mol) + 2 \times (-431\ kJ/mol) = -183\ kJ/mol.$

    The sign of bond energy terms in the calculation: when bond is broken as in reactants (H–H and Cl–Cl bonds), it will absorb heat, and the sign is positive (+); when bond is formed as in products (H–Cl bond), heat is released and sign is negative (–).

27. **(C)** To convert the same amount of carbon to CO_2, the amount of O_2 needed is simply double the amount of O_2 needed to convert same amount of carbon to CO.

28. **(E)** Atom is the basic unit which cannot be changed through chemical means, but can be changed through nuclear reaction, such as nuclear fission and fusion. Radioactive decay may also occur in which atoms of one element change to different element. However, this is not part of the Atomic Theory.

29. **(C)** is correct answer.

(A) CO_2 has a ~~bent~~ geometry — Incorrect – CO_2 has a *linear* geometry

(B) H_2O has a ~~linear~~ geometry – Incorrect – H_2O has a *bent* geometry

(C) CH_4 has a tetrahedral geometry — Correct

(D) BF_3 has a ~~pyramidal~~ geometry – Incorrect – BF_3 has a *trigonal planar* geometry

(E) NH_3 has a ~~trigonal planar~~ geometry – incorrect – NH_3 has a *trigonal pyramidal* geometry

30. **(E)** The two atoms in question have different atomic numbers (9 and 10), but same mass number (20). Therefore, they don't belong to the same element (A is incorrect), and have different number of electrons (C is incorrect). Their neutron numbers are 11 and 10 respectively based on information given (D is incorrect). They do have same mass number (20) which is the sum of protons and neutrons (E is correct answer). And finally, isomer refers to molecules which has same formula but different structure, not refers to atom (B is incorrect).

31. **(D)** This problem is about **Le Chatelier's Principle**, which states that if a chemical system at equilibrium experiences a change in concentration, temperature, volume, or partial pressure, then the equilibrium shifts to counteract the imposed change and a new equilibrium is established (**Table** 1.5).

Based on this principle, when the volume of the system is decreased, the pressure increase; and therefore, the equilibrium will shift at direction to reduce the pressure. Reactions A, B and C don't change the pressure of the system. Reaction E will shift at reverse direction, decrease in number of mole of product. Only reaction D shift at forward direction (increase amount of products).

32. **(D)** The precipitation reaction is: $AgNO_3(aq) + NaCl(aq) \rightarrow NaNO_3(aq) + AgCl(s)$

The molar ratio of reactant $AgNO_3(aq)$ and production $AgCl(s)$ in the reaction is 1:1.

The amount of $AgCl(s)$ converted to mol is (14.3 gram)/(143 gram/mol) = 0.1 mol $AgCl$.

Therefore, the amount of $AgNO_3$ in 250 mL of solution is also 0.1 mol. The molar concentration of $AgNO_3$ solution is calculated as

$$\frac{\frac{0.1 \text{ mol}}{250 \text{ mL}}}{1000 \text{mL/L}} = 0.4 \frac{\text{mol}}{\text{L}} = 0.4 \text{ M}$$

The calculation can be written with factor label method as:

$$\frac{0.1 \text{ mol}}{250 \text{ mL}} \times \frac{1000 \text{ mL}}{1 \text{ L}} = 0.4 \frac{\text{mol}}{\text{L}} = 0.4 \text{ M}$$

33. **(B)** Molarity of $C_6H_{12}O_6$ is

$$\frac{360 \text{ gram}}{180 \text{ gram/mol}} \times \frac{1}{2.0 \text{ L}} = 1.0 \frac{\text{mol}}{\text{L}} = 1.0 \text{ M}$$

34. **(E)** Referring to the solubility rules in **Table** 2.3, reactions A–D will produce precipitation $BaSO_4$, Hg_2S, $PbSO_4$, AgI. No precipitation formed in reaction E since $CaBr_2$ is soluble.

35. **(B)** is correct since HCl is the only strong electrolyte, and has greatest electricity conductivity given the concentrations of all solutions are the same. Hydrofluoric acid (C) and acetic acid (A) are weak acid. Sucrose (D) and methanol (E) are not electrolytes.

36. **(B)** Electrons in the partially filled *d* orbitals are easily excited and absorb visible light, causing the color of the transition metal complexes.

37. **(E)** Grounding the solid increases surface area, and increase dissolution rate (I is correct). Increasing temperature increases kinetic energy of molecules of solute, and speed up dissolution (II is correct). Stirring the mixture increase rate of molecular diffusion and increases rate of dissolution (III is correct).

38. **(D)** Number of mol of water molecules in 32 grams of water vapor is

(32 gram)/(18 gram/mol) = 1.78 mol.

All other choices equals to 1 mol (See **Table** 1.3 for conversion among mass, mol, and number of particles).

39. **(C)** The correct statement is "The empirical formula of $C_6H_{12}O_6$ is CH_2O".

    Butyne (A, C_4H_6) and ammonia (B, NH_3) are covalent compounds, their empirical formula are the simplest whole number ratio of all elements in the compounds, i.e. C_2H_3 and NH_3 respectively. Ionic compounds don't have recognizable molecules, and their formula is empirical formula (D is correct). Molecular formula of methane (CH_4) is already the simplest whole number ration of C and H, and its empirical formula is the same as its molecular formula (E is correct).

40. **(B)** Assuming the molar mass of the nitrogen oxide is M, the number of mol of N (molar mass 14) and O (molar mas 16) in 1 mol of the compound are:

    Number of mol of N = 0.26 x M/14, number of mol of O = 0.74 x M/16

    The molar ratio of N to O is: $\frac{0.26 \text{ x M}/14}{0.74 \text{ x M}/16} = \frac{0.26 \times 16}{0.74 \times 14} \approx \frac{1}{2}$

    Therefore, the formula of the compound is NO_2.

41. **(B)** The molar ration of Fe to Fe_2O_3 in balanced equation is 4:2 = 2:1.

    The number of mol of Fe is: (223.4 grams Fe)/(55.85 gram Fe /mol Fe) = 4 mol Fe.

    Therefore, the number of mol of Fe_2O_3 produced is 2 mol, and its mass is:

    (160 gram Fe_2O_3/mol Fe_2O_3) x (2 mol Fe_2O_3) = 320 grams Fe_2O_3.

42. **(D)** This question is about limiting reagents, follow the steps below:

    Step 1: Determine the limiting species. Since molar ratio of Na to Cl_2 is 2:1, 5 mol of Na need 2.5 mol of Cl_2 to react completely. Therefore, Cl_2 is excessive, and 0.5 mol of Cl_2 will be left after the reaction is completed.

    Step 2: Determine how many mol NaCl is produced. Production of NaCl is at a ratio of 1:1 to Na, 5 moles of Na lead to production of 5 mol of NaCl.

43. **(B)** The neutral sulfur atom (Z = 16) has 6 valence electrons. The –1 ion has 7 valence electrons.

44. **(B)** Since this reaction releases heat, increase in temperature will drive the equilibrium to proceed to the reverse direction. All other choices would drive the reaction to proceed to forward direction (**Table** 1.5).

45. **(A)** Bronsted acid donates proton in a reaction. After HS^- give away a proton (H^+), it becomes S^{2-}.

46. **(A)** After balancing the equation, $Al_2(C_2O_4)_3(s) \rightarrow Al_2O_3(s) + 3CO(g) + 3CO_2(g)$, it can be seen that the molar ration of CO to CO_2 is 1:1.

47. **(C)** At constant temperature, the volume of an ideal gas is in reverse proportion with pressure. This is described by Boyle's Law, $V_1/V_2 = P_2/P_1$. Therefore, volume will be halved if pressure is doubled at constant temperature.

48. **(C)** **Solubility product constants** (K_{sp}) are used to describe saturated solutions of ionic compounds of relatively low solubility. A saturated solution is in a state of dynamic equilibrium between the dissolved, dissociated, ionic compound and the undissolved solid.

    $M_xA_y(s) \leftrightarrow x\ M^{y+}(aq) + y\ A^{x-}(aq)$

    The general equilibrium constant (solubility product constant) for such processes can be written as:

    $K_{sp} = [M^{y+}]^x [A^{x-}]^y$.

49. **(B)** Enthalpy change of a reaction refers to change of total energy of the system (difference in potential energy between products and reactants). On an energy diagram, this is represented by the vertical distance between the energy levels of products (E) and reactants (A), which is B.

50. **(C)** The oxidation number of Cr in $Cr_2O_7^{2-}$ and Cr^{3+} are +6 and +3 respectively. Therefore, each Cr atom in $Cr_2O_7^{2-}$ must gain 3 electrons to become Cr^{3+}. This is a **reduction reaction**, and the (half) reaction equation is: $Cr_2O_7^{2-} + 14H^+ + 6e^- \rightarrow 2Cr^{3+} + 7H_2O$. (**Note**: including H^+ and H_2O is necessary to balance the equation).

51. **(C)** The reaction equation of electrolysis of KBr is: $KBr(aq) \rightarrow K(s) + Br_2(g)$.

    At the anode, electrons are "pumped out" by external battery, and the electrons are supplied by oxidation of Br^-. The half reaction equation is: $2Br^- \rightarrow Br_2 + 2e^-$.

52. **(E)** When Na_2SO_4 is added, concentration of SO_4^{2-} increase, causing the equilibrium shift to the left. This reduces the concentration of Pb^{2+}. Therefore, choice A and B are incorrect. However, if the temperature is not changed, the solubility constant K_{sp} remains unchanged (**E is correct**, and C and D are incorrect).

53. **(D)** Molar mass of NaOH is 40 gram/mol, 20 grams NaOH equals to 0.5 mol NaOH.

    To prepare 0.5 M NaOH solution which contains 0.5 mol of NaOH, the volume of the solution is

    (0.5 mol NaOH)/(0.5 mol/L) = 1 L.

    Choice A and B are incorrect since they does not follow the definition of molarity, which is the number of mol in one liter of solution.

54. **(E)** Both CH_4 and CO_2 have polar bond and nonpolar molecules due to the shape of the molecules, CH_4 is tetrahedral and CO_2 is linear. For other choices, at least one molecule either does not have polar bond or the molecule is polar.

55. **(B)** Assuming the volume of HCl is needed is V (in mL), then V x 1.5 M = 45.0 mL x 1.0 M. Solve the equation, V = 30 mL.

56. **(E)** is correct answer. Isotopes of the same elements have same number of protons, and therefore have the same atomic number and the **same positive charges** in its nucleus (E is incorrect). A-D are correct statements about isotopes.

57. **(C)** The electron configuration of Si (Z = 14) is $1s^2 2s^2 2p^6 3s^2 3p^2$ with **4 valence electrons** ($3s$ and $3p$, III is incorrect) and the p orbitals are not completely filled (I is correct). It has 2 unpaired $3p$ electrons (II is correct).

58. **(B)** At pH 11, $[H^+] = 10^{-11}$ M. Since $[H^+][OH^-] = 10^{-14}$, $[OH^-] = 10^{-14}/10^{-11} = 10^{-3}$ M.

    Alternatively, since pH + pOH = 14, pOH = 14 – pH = 14 – 11 = 3. Therefore, $[OH^-] = 10^{-3}$ M.

59. **(E)** In saturated hydrocarbon molecules which has more than 2 carbon in a chain, the two C–C bonds in C–C–C structure are at an angle, not linear. This is because for each C atom, the 4 valence electrons are hybridized (sp^3 hybridization), and the geometry of configuration for each C is tetrahedral.

60. **(A)** van der Waals force is weakest intermolecular force only significant when the molecules are nonpolar (See **Table** 1.2).

61. **(D)** is correct answer. In evaporation, temperature remains constant, although heat is needed to convert liquid substance to gaseous state (D is incorrect). All other statements are correct.

62. **(C)** In organic chemistry, orbital hybridization is an important concept as it explains the 3D configurations of molecules. Three are three types hybridization when carbon form compounds (**Table** 3.2).

In ethyne CH≡CH, both carbon have *sp* hybridization, and two electrons are filled to the *sp* hybridized orbitals. The two *sp* electrons are linear, and one form sigma bond with H and another is used to form sigma bond with another C. The 2 2*p* electron of carbon form two pi bonds with another carbon.

Table 3.2 *Orbital Hybridization and Properties*			
Type of hybridization	sp^3	sp^2	sp
Atomic orbitals used	*s, p, p, p*	*s, p, p*	*s, p*
Number of hybrid orbitals formed	4	3	2
Number of atoms bonded to the C	4	3	2
Number of sigma (σ) bonds	4	3	**1**
Number of left over *p* orbitals	0	1	2
Number of pi (π) bonds	0	1	**2**
Bonding pattern	$\begin{array}{c} \vert \\ -C- \\ \vert \end{array}$	$\begin{array}{c} \backslash \\ C= \\ / \end{array}$	=C= or –C≡
Geometry	tetrahedral	Trigonal planar	linear

63. **(E)** On a periodic table, properties which increase from left to right are electronegativity, first ionization energy, nonmetallic character, number of charge in nucleus, electron affinity. The properties decreasing from left to right include atom radius, **metallic character**.

64. **(A)** In this reaction, Cu(*s*) loses electrons to become Cu^{2+}, and is **oxidized**; N (+5) in NO_3^- gains electrons to be reduced to N (+4) in NO_2. Oxidation numbers of other atoms (H and O) remain the same before and after the reaction.

65. **(B)** One common mistake in writing formula for ionic compounds is forgetting the rule of electric charge balance. The following is a stepwise approach for writing empirical formula for iron(III) sulfate:

Table 3.3 *Write Formula of Ionic Compounds*
1. Identify cation and number of charge, Fe^{3+}. Cation always comes first.
2. Identify anion and number of charge, SO_4^{2-}.
3. Balance the electric charge. Since iron(III) has 3 positive charge and sulfate ion has 2 negative charge, the ratio of Fe^{3+} to SO_4^{2-} in the formula should be 2:3 to balance the total charge.
4. Write the empirical formula: $Fe_2(SO_4)_3$. Make sure that the numbers of Fe^{3+} and SO_4^{2-} are reduced to the simplest whole number.

66. **(C)** This is a dilution question, you need to use the law of conservation of mass as follow:
   a. Write the equation: $M_{stock}V_{stock} = M_{dilution}V_{dilution}$
   b. Determine unknown and known quantities: V_{stock} is unknown; known quantities are M_{stock} = 10.0 M, $M_{dilution}$ = 0.5 M, $V_{dilution}$ = 2 L.
   c. Solve the equation: $V_{stock} = M_{dilution}V_{dilution} / M_{stock}$ = (0.5 M) x (2 L) / (10 M) = 0.1 L = 100 mL.

67. **(C)** All radioactive decay reactions have characteristic half–lives, which are the time needed for the original radioactive isotopes to be reduced to their half of the original amount. For a radioactive substance to be reduced from 100 grams to 6.25 grams, it is reduced by 100/6.25 = 16 times, and this means that the original radioactive substance is halved 4 times; i.e. 4 half–lives have passed. Since the total time is 100 days, the half–life of this radioactive substance is 100/4 = 25 (days).

68. **(E)** is correct answer, since interaction exists between ions and ions and between molecules and ions in NaCl solution. See **Table** 1.2 for information about different bonds.

(A) is incorrect because van der Waals force exists among all molecules, polar or nonpolar. (B) is incorrect because dipolar attraction is force between molecules for molecules which has permanent polarity or nonpermanent polarity, not the covalence bond. (C) is incorrect because the bond in Cu(s) is metallic, rather than network covalence bond. (D) is incorrect since hydrogen bonds in HCl is not formed or very weak; main attraction between HCl molecules is dipolar attraction.

69. **(D)** An **amphoteric species** is a molecule or ion that can react as an acid (donate proton) and a base (accept proton). Based on this definition, H_2O in I and HS^- in II both show amphoteric property. The reactions in III are simply neutralization between a strong acid and strong base, and its reverse reaction.

70. **(D)** is correct answer, because the equation is **not balanced**, and the reaction is a **double replacement** reaction, not a single replacement reaction. The balanced equation is: $2KI + Pb(NO_3)_2 \rightarrow PbI_2 + 2KNO_3$. All other statements are correct.

101. **Correct answer: I True, II False, CE No**

Explanation: The element which has outermost configuration of $3s^1$ orbital is alkali metal and shows metallic properties (statement I is true). Metals tend to **lose** this valence electron(s) in reaction with nonmetals (Statement II is false).

102. **Correct answer: I True, II True, CE Yes**

Explanation: Heat is absorbed in an endothermic reaction (positive ΔH) since the total potential energy (enthalpy) of products is greater than that of reactant (Statements I and II are both true). And statement II is correct explanation of statement I (**Table** 3.4).

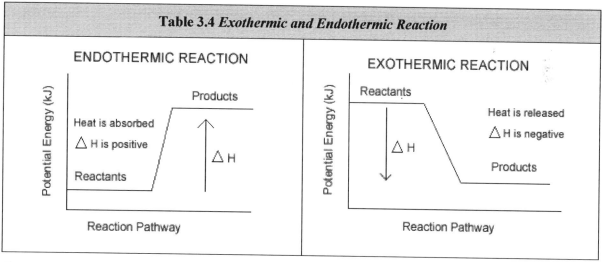

Table 3.4 *Exothermic and Endothermic Reaction*

103. **Correct answer: I True, II True, CE Yes**

Explanation: A buret is designed to measure change in volume of solution in titration accurately (Statement II is true), and normally used in volumetric titration (statement I is correct). And Statement II is appropriate explanation of statement I.

104. **Correct answer: I True, II True, CE Yes**

Explanation: Unlike voltaic cell, an electrolytic cell needs external electrical energy to drive reduction–oxidation reaction, which is nonspontaneous. Both statements are true, and statement II is appropriate explanation of statement I.

105. **Correct answer: I True, II True, CE Yes**

   Explanation: The halogens have 7 valence electrons, and need 1 electron to achieve octet state (statement II is true). When forming diatomic molecules, a covalent bond is formed between two atoms to make the number of electron on the outermost shell to be 8. Therefore, all halogens can form stable diatomic molecules (statement I is true). Statement II is appropriate explanation of statement I.

106. **Correct answer: I True, II False, CE No**

   Explanation: 3 kilograms equal 3000 grams (statement I is correct) because the prefix kilo– means "one **thousand**", not "one **thousandth**" (statement II is incorrect).

107. **Correct answer: I True, II True, CE Yes**

   Explanation: When temperature of a gas increases, the average kinetic energy of the molecules increase (statement II is true), and the pressure increases. If the pressure remains the same, the volume of the gas must increase to offset the increase in pressure caused by increase in kinetic energy (statement I is true). Statement II is appropriate explanation of statement I.

108. **Correct answer: I True, II False, CE No**

   Explanation: A catalyst can speed up a reaction by lowering the activation energy of both reactant and product (statement I is true). However, catalyst does not change the enthalpy (heat or potential energy) of either reactant or product, and therefore, change enthalpy of the reaction (ΔH) is not affected by catalyst (Statement II is false).

109. **Correct answer: I True, II True, CE Yes**

   Explanation: At low temperature and high pressure, the distance between molecules are close enough to allow the interaction between molecules to be significant, and cannot be neglected. Both statements are correct and statement II is appropriate explanation of statement I.

110. **Correct answer: I True, II True, CE No**

   Explanation: The lighter the gas molecule, the faster its effusion rate (effusion rate is inversely proportional to the square root of molar mass). Therefore diatomic nitrogen effuses faster than diatomic oxygen (Statement I is true). However, the type of bond of diatomic N_2 and O_2 has nothing with the effusion rate. Although statement II is correct, but not correct explanation of statement I.

111. **Correct answer: I True, II False, CE No**

   Explanation: A chemical compound is a pure chemical substance consisting of two or more different chemical elements that can be separated into simpler substances by chemical reactions, and propane (C_3H_8) is an organic compound (statement I is correct). The bonds in a compound can be broken by chemical reactions (Statement II is false).

112. **Correct answer: I True, II True, CE Yes**

   Explanation: When a mixture of two liquid is heated, the one with lower boiling point evaporate first, and its vapor is condensed and collected. This process for separating liquid mixture is called distillation. Both statements are true, and statement II is appropriate explanation of statement I.

113. **Correct answer: I False, II True, CE No**

   Explanation: Isotopes of an element must have the same atomic number (number of protons, statement I is false), but different number of neutrons (statement II is true).

114. **Correct answer: I True, II True, CE Yes**

   Explanation: The solution of water has lower melting point than pure water, since the concentration of water in a solution is lower than pure water. This phenomenon is called melting point depression.

(Statement I is true). This is caused by lower vapor pressure due to addition of solute (**Figure** 3.2) Statement II is appropriate explanation of statement I.

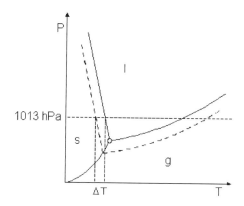

Figure 3.2 Phase diagram of pure solvent and solution. Dotted lines are new boundaries due to addition of solute.

115. **Correct answer: I True, II True, CE Yes**

Explanation: Radioactivity has a lot of beneficial applications in many areas. Both statements are correct, and statement II is appropriate explanation of statement I.

SAT II Chemistry

Practice Test 4

Answer Sheet

Part A and C: Determine the correct answer. Blacken the oval of your choice completely with a No. 2 pencil.

1	Ⓐ Ⓑ Ⓒ Ⓓ Ⓔ	25	Ⓐ Ⓑ Ⓒ Ⓓ Ⓔ	49	Ⓐ Ⓑ Ⓒ Ⓓ Ⓔ
2	Ⓐ Ⓑ Ⓒ Ⓓ Ⓔ	26	Ⓐ Ⓑ Ⓒ Ⓓ Ⓔ	50	Ⓐ Ⓑ Ⓒ Ⓓ Ⓔ
3	Ⓐ Ⓑ Ⓒ Ⓓ Ⓔ	27	Ⓐ Ⓑ Ⓒ Ⓓ Ⓔ	51	Ⓐ Ⓑ Ⓒ Ⓓ Ⓔ
4	Ⓐ Ⓑ Ⓒ Ⓓ Ⓔ	28	Ⓐ Ⓑ Ⓒ Ⓓ Ⓔ	52	Ⓐ Ⓑ Ⓒ Ⓓ Ⓔ
5	Ⓐ Ⓑ Ⓒ Ⓓ Ⓔ	29	Ⓐ Ⓑ Ⓒ Ⓓ Ⓔ	53	Ⓐ Ⓑ Ⓒ Ⓓ Ⓔ
6	Ⓐ Ⓑ Ⓒ Ⓓ Ⓔ	30	Ⓐ Ⓑ Ⓒ Ⓓ Ⓔ	54	Ⓐ Ⓑ Ⓒ Ⓓ Ⓔ
7	Ⓐ Ⓑ Ⓒ Ⓓ Ⓔ	31	Ⓐ Ⓑ Ⓒ Ⓓ Ⓔ	55	Ⓐ Ⓑ Ⓒ Ⓓ Ⓔ
8	Ⓐ Ⓑ Ⓒ Ⓓ Ⓔ	32	Ⓐ Ⓑ Ⓒ Ⓓ Ⓔ	56	Ⓐ Ⓑ Ⓒ Ⓓ Ⓔ
9	Ⓐ Ⓑ Ⓒ Ⓓ Ⓔ	33	Ⓐ Ⓑ Ⓒ Ⓓ Ⓔ	57	Ⓐ Ⓑ Ⓒ Ⓓ Ⓔ
10	Ⓐ Ⓑ Ⓒ Ⓓ Ⓔ	34	Ⓐ Ⓑ Ⓒ Ⓓ Ⓔ	58	Ⓐ Ⓑ Ⓒ Ⓓ Ⓔ
11	Ⓐ Ⓑ Ⓒ Ⓓ Ⓔ	35	Ⓐ Ⓑ Ⓒ Ⓓ Ⓔ	59	Ⓐ Ⓑ Ⓒ Ⓓ Ⓔ
12	Ⓐ Ⓑ Ⓒ Ⓓ Ⓔ	36	Ⓐ Ⓑ Ⓒ Ⓓ Ⓔ	60	Ⓐ Ⓑ Ⓒ Ⓓ Ⓔ
13	Ⓐ Ⓑ Ⓒ Ⓓ Ⓔ	37	Ⓐ Ⓑ Ⓒ Ⓓ Ⓔ	61	Ⓐ Ⓑ Ⓒ Ⓓ Ⓔ
14	Ⓐ Ⓑ Ⓒ Ⓓ Ⓔ	38	Ⓐ Ⓑ Ⓒ Ⓓ Ⓔ	62	Ⓐ Ⓑ Ⓒ Ⓓ Ⓔ
15	Ⓐ Ⓑ Ⓒ Ⓓ Ⓔ	39	Ⓐ Ⓑ Ⓒ Ⓓ Ⓔ	63	Ⓐ Ⓑ Ⓒ Ⓓ Ⓔ
16	Ⓐ Ⓑ Ⓒ Ⓓ Ⓔ	40	Ⓐ Ⓑ Ⓒ Ⓓ Ⓔ	64	Ⓐ Ⓑ Ⓒ Ⓓ Ⓔ
17	Ⓐ Ⓑ Ⓒ Ⓓ Ⓔ	41	Ⓐ Ⓑ Ⓒ Ⓓ Ⓔ	65	Ⓐ Ⓑ Ⓒ Ⓓ Ⓔ
18	Ⓐ Ⓑ Ⓒ Ⓓ Ⓔ	42	Ⓐ Ⓑ Ⓒ Ⓓ Ⓔ	66	Ⓐ Ⓑ Ⓒ Ⓓ Ⓔ
19	Ⓐ Ⓑ Ⓒ Ⓓ Ⓔ	43	Ⓐ Ⓑ Ⓒ Ⓓ Ⓔ	67	Ⓐ Ⓑ Ⓒ Ⓓ Ⓔ
20	Ⓐ Ⓑ Ⓒ Ⓓ Ⓔ	44	Ⓐ Ⓑ Ⓒ Ⓓ Ⓔ	68	Ⓐ Ⓑ Ⓒ Ⓓ Ⓔ
21	Ⓐ Ⓑ Ⓒ Ⓓ Ⓔ	45	Ⓐ Ⓑ Ⓒ Ⓓ Ⓔ	69	Ⓐ Ⓑ Ⓒ Ⓓ Ⓔ
22	Ⓐ Ⓑ Ⓒ Ⓓ Ⓔ	46	Ⓐ Ⓑ Ⓒ Ⓓ Ⓔ	70	Ⓐ Ⓑ Ⓒ Ⓓ Ⓔ
23	Ⓐ Ⓑ Ⓒ Ⓓ Ⓔ	47	Ⓐ Ⓑ Ⓒ Ⓓ Ⓔ	71	Ⓐ Ⓑ Ⓒ Ⓓ Ⓔ
24	Ⓐ Ⓑ Ⓒ Ⓓ Ⓔ	48	Ⓐ Ⓑ Ⓒ Ⓓ Ⓔ	72	Ⓐ Ⓑ Ⓒ Ⓓ Ⓔ

Part B: On the actual Chemistry Test, the following type of question must be answered on a special section (labeled "Chemistry") at the lower left–hand corner of your answer sheet. These questions will be numbered beginning with 101 and must be answered according to the directions.

PART B	I		II		CE
101	T	F	T	F	◯
102	T	F	T	F	◯
103	T	F	T	F	◯
104	T	F	T	F	◯
105	T	F	T	F	◯
106	T	F	T	F	◯
107	T	F	T	F	◯
108	T	F	T	F	◯
109	T	F	T	F	◯
110	T	F	T	F	◯
111	T	F	T	F	◯
112	T	F	T	F	◯
113	T	F	T	F	◯
114	T	F	T	F	◯
115	T	F	T	F	◯
116	T	F	T	F	◯

Periodic Table of Elements

Material in this table may be useful in answering the questions in this examination

1 H 1.0079																		2 He 4.0026
3 Li 6.941	4 Be 9.012											5 B 10.811	6 C 12.011	7 N 14.007	8 O 16.00	9 F 19.00	10 Ne 20.179
11 Na 22.99	12 Mg 24.30											13 Al 26.98	14 Si 28.09	15 P 30.974	16 S 32.06	17 Cl 35.453	18 Ar 39.948
19 K 39.01	20 Ca 40.48	21 Sc 44.96	22 Ti 47.90	23 V 50.94	24 Cr 52.00	25 Mn 54.938	26 Fe 55.85	27 Co 58.93	28 Ni 58.69	29 Cu 63.55	30 Zn 65.39	31 Ga 69.72	32 Ge 72.59	33 As 74.92	34 Se 78.96	35 Br 79.90	36 Kr 83.80
37 Rb 85.47	38 Sr 87.62	39 Y 88.91	40 Zr 91.22	41 Nb 92.91	42 Mo 95.94	43 Tc (98)	44 Ru 101.1	45 Rh 102.91	46 Pd 106.42	47 Ag 107.87	48 Cd 112.41	49 In 114.82	50 Sn 118.71	51 Sb 121.75	52 Te 127.60	53 I 126.91	54 Xe 131.29
55 Cs 132.91	56 Ba 137.33	57 *La 138.91	72 Hf 178.49	73 Ta 180.95	74 W 183.85	75 Re 186.21	76 Os 190.2	77 Ir 192.2	78 Pt 195.08	79 Au 196.97	80 Hg 200.59	81 Tl 204.38	82 Pb 207.2	83 Bi 208.98	84 Po (209)	85 At (210)	86 Rn (222)
87 Fr (223)	88 Ra 226.02	89 Ac 227.03	104 Rf (261)	105 Db (262)	106 Sg (266)	107 Bh (264)	108 Hs (277)	109 Mt (268)	110 Ds (271)	111 Rg (272)	112 (277)						

*Lanthanide Series	58 Ce 140.12	59 Pr 140.91	60 Nd 144.24	61 Pm (145)	62 Sm 150.4	63 Eu 151.97	64 Gd 157.25	65 Tb 158.93	66 Dy 162.50	67 Ho 164.93	68 Er 167.26	69 Tm 168.93	70 Yb 173.04	71 Lu 174.97
Actinide Series	90 Th 232.04	91 Pa 231.04	92 U 238.03	93 Np 237.05	94 Pu (244)	95 Am (243)	96 Cm (247)	97 Bk (247)	98 Cf (251)	99 Es (252)	100 Fm (257)	101 Md (258)	102 No (259)	103 Lr (260)

Note: For all questions involving solutions, assume that the solvent is water unless otherwise stated.
Reminder: You may not use a calculator in this test!

Throughout the test the following symbols have the definitions specified unless otherwise noted.

H = enthalpy	atm = atmosphere(s)
M = molar	g = gram(s)
n = number of moles	J = joule(s)
P = pressure	kJ = kilojoule(s)
R = molar gas constant	L = liter(s)
S = entropy	mL = milliliter(s)
T = temperature	mm = millimeter(s)
V = volume	mol = mole(s)
	V = volt(s)

Chemistry Subject Practice Test 4

Part A

Directions for Classification Questions

Each set of lettered choices below refers to the numbered statements or questions immediately following it. Select the one lettered choice that best fits each statement or answers each question and then fill in the corresponding circle on the answer sheet. A choice may be used once, more than once, or not at all in each set.

Questions 1 – 4 refer to the following

    (A) Boiling point
    (B) Rate of reaction
    (C) Molecular mass
    (D) Molarity
    (E) Density

1. Can be expressed as grams per milliliter

2. Can be expressed as grams per mole

3. Does not vary with changes of temperature and pressure

4. Is a quantity necessary to determine the molecular formula of a compound

Questions 5 – 8 refer to the following

    (A) Precipitation
    (B) Oxidation–reduction
    (C) Distillation
    (D) Hydration
    (E) Neutralization

5. Electrolysis of water to form hydrogen and oxygen gases

6. Reaction of silver ion with chloride ion in water solution

7. Reaction between an acid and a base

8. Reaction iron filings with powdered sulfur

Questions 9 – 12 refer to the following substances

    (A) H_2
    (B) CO_2
    (C) H_2O
    (D) NaCl
    (E) CH_2CH_2

9. Contains just one sigma bond

10. Has a bond formed from the transfer of electrons

11. Has an atom that is *sp* hybridized

12. Is a polar molecule

Questions 13 – 16 refer to the following elements

    (A) F
    (B) Li
    (C) Fe
    (D) He
    (E) Si

13. Shows both the properties of both metals and non–metals

14. Has the greatest ionization energy

15. Has the greatest electronegativity

16. Has colored salts that will produce colored aqueous solutions

GO ON TO THE NEXT PAGE

Questions 17 – 21 refer to the following compounds

    (A) $CaCO_3$
    (B) $AgNO_3$
    (C) KCl
    (D) NH_3
    (E) HCl

17. Forms a white precipitate when added to a solution of NaCl

18. Will form a coordinate covalent bond with a hydronium ion

19. Is a strong acid in solution

20. Release a colorless gas with addition of diluted acid

21. A product of a neutralization of a strong acid with a strong base

Questions 22 – 25 refer to the following

    (A) Exothermic
    (B) Endothermic
    (C) At equilibrium
    (D) Basic
    (E) Supersaturated

22. Condition that exists in acetic acid solution

23. Combustion of magnesium in air

24. A reaction must absorb heat to proceed

25. Concentrated sulfuric acid is dissolved in water

GO ON TO THE NEXT PAGE

PLEASE GO TO THE SPECIAL SECTION AT THE LOWER LEFT–HAND CORNER OF PAGE 2 OF YOUR ANSWER SHEET LABELED CHEMISTRY AND ANSWER QUESTIONS 101–115 ACCORDING TO THE FOLLOWING DIRECTIONS.

Part B

Directions for Relationship Analysis Questions
Each question below consists of two statements, I in the left–hand column and II in the right–hand column. For each question, determine whether statement I is true or false and whether statement II is true or false and fill in the corresponding T or F circles on your answer sheet. *Fill in circle CE only if statement II is a correct explanation of the true statement I.

EXAMPLES:

I		II
EX1. The nucleus in an atom has a positive charge.	BECAUSE	Proton has positive charge, neutron has no charge.

SAMPLE ANSWERS

	I	II	CE
EX1	● Ⓕ	● Ⓕ	●

	I		II
101.	Some alpha particles shot at a thin metal foil are reflected back toward the source	BECAUSE	Alpha particles shot at thin metal foil sometimes approach the nuclei of the metal atoms head–on bond and are thus repelled.
102.	The electron configuration for nitrogen is $1s^2 2s^2 2p^5$	BECAUSE	nitrogen has five valence electrons.
103.	A molecule of H_2O is linear	BECAUSE	the oxygen atoms in H_2O molecules are *sp* hybridized.
104.	AgCl dissolves in water	BECAUSE	all chloride salts are soluble in water.
105.	CCl_4 is a polar molecule	BECAUSE	C–Cl bonds in CCl_4 molecules are polar.
106.	Molten potassium chloride is a good electrical conductor	BECAUSE	the melting process frees the ions in KCl(s) from their fixed position in the crystal lattice.

GO ON TO THE NEXT PAGE ▷

107. HCl is an Arrhenius acid in solution   **BECAUSE**   HCl will yield hydronium ions as the only positive ions.

108. Adding more reactants will speed up a reaction   **BECAUSE**   the reactants will collide less frequently.

109. $Cu^{2+} + 2e^- \rightarrow Cu$ is a correctly balanced oxidation reaction   **BECAUSE** $Cu^{2+} + 2e^- \rightarrow Cu$ correctly demonstrates conservation of mass and conservation of charge

110. If the volume on a buret can be read to the nearest 0.01 mL, then the volume of exactly 15 mL of a solution released from the buret should be recorded as 15.00 mL   **BECAUSE**   it is standard practice to record data to two decimal places.

111. Fluorine has the highest value for electronegativity   **BECAUSE**   fluorine has the greatest attraction for electrons.

112. Covalent bonds are broken when sugars are dissolve in water   **BECAUSE**   in molecules with covalent bonds, electrons are shared by two or more atoms.

113. Sodium has a smaller atomic radius than chlorine   **BECAUSE**   a sodium atom does not have as many valence electrons as a chlorine atom does.

114. Elemental sodium is a good reducing reagent.   **BECAUSE**   sodium atoms readily give up their valence electrons.

115. A 0.1 M NaCl(*aq*) solution will freeze at a temperature below 0°C   **BECAUSE**   as a solute is added to a solvent, the boiling point and freezing point decrease.

GO ON TO THE NEXT PAGE

Part C

Directions for Five–Choice Completion Questions
Each of the questions or incomplete statements below is followed by five suggested answers or completions. Select the one that is best in each case and then fill in the corresponding circle on the answer sheet.

26. Which of the following properties generally decreases when moving from left to right across a period (row) of the periodic table?

    (A) Reactivity
    (B) Atomic radius
    (C) Electron affinity
    (D) Ionization energy
    (E) Number of valence electrons

27. Which of the following solution is the best electrolyte?

    (A) 0.2 M of weak acid 1, $K_a = 1.8 \times 10^{-7}$
    (B) 0.5 M of weak acid 2, $K_a = 2.3 \times 10^{-8}$
    (C) 1.5 M of weak acid 3, $K_a = 6.1 \times 10^{-10}$
    (D) 1.0 M of weak acid 4, $K_a = 3.7 \times 10^{-3}$
    (E) 0.9 M of weak acid 5, $K_a = 5.9 \times 10^{-4}$

28. Which of the following is true about a solution that has a concentration of OH^- of 1.0×10^{-8} M?

    (A) The pH is 8 and the solution is basic
    (B) The concentration of H^+ is 1.0×10^{-6} M and the solution is basic
    (C) The pH is 6 and the solution is acidic
    (D) The pH is 4 the solution is acidic
    (E) The concentration of H^+ is 1.0×10^{-7} M and the solution is neutral

29. What will be the products of the following double replacement reaction?

    $$(NH_4)_3PO_4 + Ba(NO_3)_2 \rightarrow$$

    (A) Ammonium nitrate and barium nitrate
    (B) Barium nitrate and ammonium phosphate
    (C) Barium phosphate and sodium nitrate
    (D) Ammonium nitrate and barium phosphate
    (E) Ammonium nitrate and barium nitrate

30. The diatomic molecular bromine, Br_2, is a nonpolar substance, but exists as liquid at room temperature. Which of the following explain this phenomenon?

    (A) The Br_2 molecules form temporary dipoles with partial charge.
    (B) The Br_2 molecules can transfer electron, forming ions with opposite charges
    (C) The Br_2 molecules form positive ions that are held together by free electrons
    (D) Hydrogen bonds are formed between Br_2 molecules
    (E) The Br_2 molecules share electrons, forming bonds.

31. A student performs a titration using 1.00 M NaOH to find the unknown molarity of 100 mL HCl. The student records the data as shown below. What is the molarity of the solution of HCl?

    Base: final buret reading 21.05 mL
    Base: initial buret reading 6.05 mL

    (A) 0.05 M
    (B) 0.10 M
    (C) 0.15 M
    (D) 0.25 M
    (E) 1.50 M

32. Which of the following descriptions regarding alkanes is NOT correct?

    (A) Only two elements, C and H, are found in their molecules
    (B) The simplest member is methane, CH_4
    (C) They are all saturated molecules
    (D) They are nonpolar molecules
    (E) They are soluble in water

GO ON TO THE NEXT PAGE

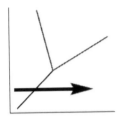

33. Which of the following terms is the correct representation of the process shown by the arrow on the phase diagram above?

    (A) Evaporation
    (B) Deposition
    (C) Condensation
    (D) Freezing
    (E) Sublimation

34. $2NO(g) + O_2(g) \rightleftharpoons 2NO_2 + 114$ kilojoules

    If 0.5 mole of NO_2 is formed in the reaction above, how much heat is released?

    (A) 228
    (B) 114 kJ
    (C) 57 kJ
    (D) 28.5 kJ
    (E) 14.3 kJ

35. What is the percent composition of oxygen in glucose, $C_6H_{12}O_6$ (molar mass = 180)?

    (A) 20%
    (B) 38%
    (C) 42%
    (D) 53%
    (E) 81%

36. Which of the following has the greatest affinity for electrons?

    (A) F
    (B) Br
    (C) C
    (D) Fe
    (E) Na

37. At STP, a pure gas sample has a volume of 11.2 L and a mass of 22.0 grams. This gas is most likely

    (A) CO_2
    (B) O_2
    (C) N_2
    (D) H_2
    (E) NH_3

38. What is the mass–action expression (equilibrium constant) for the following reaction at equilibrium?

    $$2X(aq) + Y(s) \rightleftharpoons 3V(aq) + 2W(s)$$

    (A) $\dfrac{[V]^3[W]^2}{[X]^2[Y]}$

    (B) $\dfrac{[V][W]}{[X][Y]}$

    (C) $\dfrac{[V]}{[X]}$

    (D) $\dfrac{[V]^3}{[X]^2}$

    (E) $\dfrac{3[V]}{2[X]}$

39. Which of the following has the strongest carbon – carbon bond?

    (A) C_2H_2
    (B) C_2H_3Cl
    (C) C_2H_4
    (D) C_2H_5Cl
    (E) C_2H_6

40. A student dissolved 58.5 grams of NaCl in pure water 1.5 L of solution. What is the molarity of this solution?

    (A) 1.5 M
    (B) 1.0 M
    (C) 0.67 M
    (D) 0.33 M
    (E) 0.15 M

GO ON TO THE NEXT PAGE

41. How many atoms are represented in a formula of $Cr(NH_3)_5SO_4Br$?

(A) 6
(B) 8
(C) 12
(D) 23
(E) 27

42. A gas has a volume of 10 liter at 50°C and 450 mmHg. What is its volume at STP?

(A) 10 x (0/50) x (450/760)

(B) 10 x (0/50) x (760/450)

(C) 10 x (273/323) x (450/760)

(D) 10 x (273/323) x (760/450)

(E) 10 x (323/273) x (760/450)

43. The table below lists concentration and pH of three acid/base solutions prepared from pure substances.

Solution	Concentration (mol/L)	pH
I	0.1	14
II	0.1	2.9
III	0.01	2.0

Which of the following is correct regarding which solution is prepared with weak acid?

(A) I only
(B) II only
(C) III only
(D) I and II only
(E) II and III only

44. Which of the following systems at equilibrium will NOT be influenced by a change in pressure?

(A) $O_2(g) \rightleftharpoons O_3(g)$
(B) $N_2(g) + H_2(g) \rightleftharpoons NH_3(g)$
(C) $NO_2(g) \rightleftharpoons N_2O_4(g)$
(D) $H_2(g) + I_2(g) \rightleftharpoons HI(g)$
(E) $CO(g) + O_2(g) \rightleftharpoons CO_2(g)$

45. Elemental analysis of an unknown hydrocarbon indicates the mass ratio of carbon to hydrogen is 6:1. Which of the following is the most likely identity of this hydrocarbon?

(A) Methane, CH_4
(B) Ethane, C_2H_6
(C) Propane, C_3H_8
(D) Ethene, C_2H_4
(E) Ethyne, C_2H_2

46. $...C_3H_8O + ...O_2 \rightarrow ...CO_2 + ...H_2O$

The equation above represents the complete combustion of propanol. When this equation is balanced and all coefficients are reduced to lowest whole–number terms, the coefficient for O_2 is

(A) 6
(B) 8
(C) 9
(D) 12
(E) 18

47. Which of the following salts works best for melting an ice sheet on a sidewalk?

(A) NaCl
(B) $CaCl_2$
(C) KBr
(D) $NaNO_3$
(E) $NaC_2H_3O_2$

48. Given the photosynthesis reaction in plants:

$6CO_2 + 6H_2O \rightarrow C_6H_{12}O_6 + 6O_2$

If 96 grams of oxygen are produced by the plant, how many grams $C_6H_{12}O_6$ (molar mass = 180) can be made?

(A) 90 grams
(B) 192 grams
(C) 180 grams
(D) 540 grams
(E) 1080 grams

GO ON TO THE NEXT PAGE

49. Which of the following will NOT be changed by the addition of a catalyst to a reaction which eventually reaches equilibrium?

   I.   The equilibrium constant
   II.   The rate at which the reaction reaches equilibrium
   III.   The potential energy of the products

(A) I only
(B) II only
(C) I and III only
(D) II and III only
(E) I, II and III

50. $Pb(s) + S(s) \rightarrow PbS(s)$

If 20.7 grams of lead (mass number 207) and 6.4 grams of sulfur (mass number 32) are mixed and react as represented above, which of the following is correct?

(A) Lead is in excess by 14.3 grams
(B) Lead is in excess by 10.35 grams
(C) Sulfur is in excess by 3.2 grams
(D) Sulfur is in excess by 1.6 grams
(E) Neither lead nor sulfur is in excess

51. The kinetic molecular theory explains the behavior of

(A) Gas only
(B) Liquid only
(C) Solid only
(D) Gas and liquid only
(E) Gas, liquid and solid

52. A student is performing an experiment where a blue salt is being heated to dryness in order to determine the percent of water in the salt. The following equipment are necessary EXCEPT

(A) An analytical balance
(B) A Bunsen burner
(C) A tong
(D) A wire gauze
(E) A thermometer

53. For the following reaction:

$Fe_2O_3(s) + CO(g) \rightarrow Fe(s) + CO_2(g)$

Is allowed to completely react with 56 grams of CO, how many moles of iron, Fe, are produced?

(A) 0.7 (B) 1.3 (C) 2.0 (D) 2.7 (E) 6.0

54. At STP, one containers has 4 grams of $H_2(g)$ and another has 4 grams of $He(g)$, which of the following statements is true?

(A) The hydrogen gas occupies 44.8 liters of volume and the helium gas will contain 6.02×10^{23} molecules
(B) The hydrogen gas occupies 22.4 liters and the helium gas contains 3.02×10^{23} molecules
(C) The hydrogen gas occupies 44.8 liters and the helium gas contains 1.204×10^{24} molecules
(D) The helium gas occupies 44.8 liters and the hydrogen gas contains 6.02×10^{23} molecules
(E) None of the above statements is correct

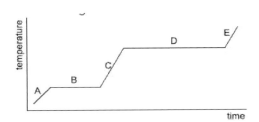

55. The diagram shows a solid being heated from below its freezing point. Which line segment shows the gas and the liquid phases existing at the same time?

(A) A (B) B (C) C (D) D (E) E

GO ON TO THE NEXT PAGE

56. $Ce^{3+} + Pb \rightarrow Ce + Pb^{4+}$

What is the coefficient of Pb after the reduction–oxidation reaction above is balanced and all coefficients are reduced to lowest whole number?

(A) 2 (B) 3 (C) 4 (D) 6 (E) 8

57. Which of the following statements is correct regarding molecular geometries?

   I. CH_4 is tetrahedral
   II. H_2O is linear
   III. $BeCl_2$ is linear

(A) I only
(B) II only
(C) III only
(D) I and III only
(E) I, II and III

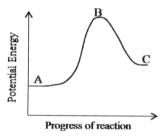

58. The diagram above shows how potential energy changes during a chemical reaction. Which aspect of the graph provides the best evidence that the reaction is endothermic?

(A) The potential energy at A
(B) The potential energy at C
(C) The energy difference between A and B
(D) The energy difference between A and C
(E) The energy difference between B and C

59. Which of the following compounds is correctly named?

(A) CO—monocarbon monoxide
(B) CaF_2—calcium difluoride
(C) CCl_4—carbon tetrachloride
(D) PCl_3—potassium trichloride
(E) TiF_4—tin(IV) fluoride

60. Which of the following gases can be dried with NaOH?

(A) CO_2
(B) SO_2
(C) HBr
(D) O_2
(E) HCl

61. Which type of intermolecular forces best explain why CH_4 is a gas, while C_8H_{18} is a liquid and $C_{20}H_{42}$ is a solid at room temperature?

   I. Hydrogen bonding
   II. van der Waals (dispersion) force
   III. dipole–dipole interaction

(A) I only
(B) II only
(C) III only
(D) I and II only
(E) I, II, and III

62. Which of the gases listed below would NOT be collected via water displacement?

(A) CO_2
(B) CH_4
(C) O_2
(D) NH_3
(E) H_2

GO ON TO THE NEXT PAGE

63. Which scientist and discovery are not correctly paired?

(A) Millikan / neutron
(B) Rutherford / nucleus
(C) Charles / relationship between temperature and volume
(D) Curie / radioactivity
(E) Mendeleyev / periodic table

64. Which of the following processes causes an increase in entropy?

  I. Dissolving a salt into water
  II. Sublimation
  III. Heating up a liquid

(A) I only
(B) I and II only
(C) II and III only
(D) I and III only
(E) I, II, and III

65. Which of the following will always decrease the volume of a gas?

  I. Decrease the pressure with temperature held constant
  II. Increase the pressure and decrease the temperature
  III. Increase both pressure and temperature simultaneously

(A) I only
(B) II only
(C) I and II only
(D) II and III only
(E) I, II, and III

66. $CH_4(g) + 2O_2(g) \rightarrow CO_2(g) + H_2O(g) + 800$ kJ

In the reaction represented above, 32 grams of oxygen is consumed to combust methane, how much heat is produced?

(A) 200 kJ
(B) 400 kJ
(C) 800 kJ
(D) 1200 kJ
(E) 1600 kJ

67. A gas is confined in the manometer as shown below. The stopcock is then opened and the highest level of mercury inside the tube moved to a level that is 18 cm above its lowest level. What is the pressure of the gas?

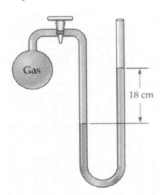

(A) 180 mmHg
(B) 360 mmHg
(C) 580 mmHg
(D) 940 mmHg
(E) The pressure cannot be determined

GO ON TO THE NEXT PAGE

68. Which of the following statements best describes the density and rate of effusion of the following gases under the same condition of pressure and temperature?

NO_2 H_2 Kr Xe F_2

(A) Fluorine has the lowest density and the lowest rate of effusion
(B) Xenon has the greatest rate of effusion and the lowest density
(C) Krypton has the lowest density and the greatest rate of effusion
(D) Hydrogen has the greatest rate of effusion and the lowest density
(E) Nitrogen dioxide has the highest density and the greatest rate of effusion

69. According to quantum mechanics, correct statements concerning the electron in the hydrogen atom include which of the following?

   I. It moves in a definite orbit around the nucleus
   II. It is associated with definite energy levels
   III. It occupies a fixed position in space with reference to the nucleus

(A) I only
(B) II only
(C) I and III only
(D) II and III only
(E) I, II and III

70. In a multistep chemical process, the rate–limiting step is the step in the reaction with the

(A) Highest activation energy and fastest reaction rate
(B) Highest activation energy and slowest reaction rate
(C) Lowest activation energy and fastest reaction rate
(D) Lowest activation energy and slowest reaction rate
(E) Greatest concentration of the reactants and products

STOP!

If you finish before time is called, you may check your work on this section only. Do not turn to any other section in the test.

Practice Test 4 Answers

PART A and C						PART B	
#	Answer	#	Answer	#	Answer	#	Answer
1	E	25	A	49	C	101	True, True, Yes
2	C	26	B	50	C	102	False, True, No
3	C	27	D	51	E	103	False, False, No
4	C	28	C	52	E	104	False, False, No
5	B	29	D	53	B	105	False, True, No
6	A	30	A	54	A	106	True, True, Yes
7	E	31	C	55	D	107	True, True, Yes
8	B	32	E	56	B	108	True, False, No
9	A	33	E	57	D	109	False, True, No
10	D	34	D	58	D	110	True, False, No
11	B	35	D	59	C	111	True, True, Yes
12	C	36	A	60	D	112	False, True, No
13	E	37	A	61	B	113	False, True, No
14	D	38	D	62	D	114	True, True, Yes
15	A	39	A	63	A	115	True, False, No
16	C	40	C	64	E	116	
17	B	41	E	65	B		
18	D	42	C	66	B		
19	E	43	B	67	D		
20	A	44	D	68	D		
21	C	45	D	69	B		
22	C	46	C	70	B		
23	A	47	B	71			
24	B	48	A	72			

Calculation of the raw score

The number of correct answers: _____ = No. of correct

The number of wrong answers: _____ = No. of wrong

Raw score = No. of correct – No. of wrong x ¼ = _____

Score Conversion Table

Raw Score	Scaled Score	Raw Score	Scaled Score	Raw Score	Scaled Score
80	800	49	600	18	420
79	800	48	590	17	410
78	790	47	590	16	410
77	780	46	580	15	400
76	770	45	580	14	390
75	770	44	570	13	390
74	760	43	560	12	380
73	760	42	560	11	370
72	750	41	550	10	360
71	740	40	550	9	360
70	740	39	540	8	350
69	730	38	540	7	350
68	730	37	530	6	340
67	720	36	520	5	340
66	710	35	520	4	330
65	700	34	510	3	330
64	700	33	500	2	320
63	690	32	500	1	320
62	680	31	490	0	310
61	680	30	490	−1	310
60	670	29	480	−2	300
59	660	28	480	−3	300
58	660	27	470	−4	290
57	650	26	470	−5	280
56	640	25	460	−6	280
55	640	24	450	−7	270
54	630	23	450	−8	270
53	620	22	440	−9	260
52	620	21	440	−10	260
51	610	20	430		
50	600	19	420		

Explanations: Practice Test 4

1. **(E)** Density of a substance is its mass in unit volume. Some commonly used density unit include grams per milliliter (g/mL) for both liquid and solid.

2. **(C)** Molecular mass of a substance is the mass of a mole of the substance. Molecular mass is usually expressed as **grams per mole** (g/mol).

3. **(C)** Molecular mass does not change with temperature and pressure as defined above. Boiling point of a liquid varies with pressure of surrounding air. Rate of reaction might change with either temperature and/or pressure. Molarity and density might change when the volume of the substance/solution changes, and the volume of solution substance might change with temperature and/or pressure.

4. **(C)** Molecular mass of a pure compound equals to the sum of molar mass of all atoms which make up the molecule of the compound. This provides important information in determining the formula of that compound. However, molecular mass itself cannot determine the formula of a compound, the relative abundance or molar ratio of atoms in the molecules is also necessary to determine the formula of the compound.

5. **(B)** In hydrolysis of water, **reduction reaction** occurs on cathode and **oxidation reaction** occurs on anode as follow:

    Cathode (reduction): $2 H_2O(l) + 2e^- \rightarrow H_2(g) + 2 OH^- (aq)$
    Anode (oxidation): $4 OH^- (aq) \rightarrow O_2(g) + 2 H_2O(l) + 4e^-$

    Combined reaction: $2H_2O(l) \rightarrow 2H_2(g) + O_2(g)$

6. **(A)** The product of the reaction between silver ion (Ag^+) and chlorine ion (Cl^-) is an insoluble salt, silver chloride (AgCl). This is a **precipitation reaction**.

7. **(E)** As defined in chemistry, neutralization is a reaction between an acid and a base, in which hydrogen ion (H^+) and hydroxide ion (OH^-) combine to form water.

8. **(B)** The reaction between iron (Fe) and sulfur (S) produces iron sulfide (FeS or Fe_2S_3). In this reaction, Fe loses electrons (oxidation) and S gains electrons (reduction).

9. **(A)** Hydrogen atom only has one $1s$ electron, the covalence bond between two hydrogen atoms is formed through overlapping of the two $1s$ electrons of the two H atoms. Therefore, there is only one covalence bond in diatomic H_2 molecule, and it's an s–s sigma bond (H:H). See **Table** 1.13 for details of valence bond theory.

10. **(D)** **Ionic bond** forms when electron is transferred from one atom to another. The atom which loses electron has positive charge, and the atom which gains electron has negative charge. The electrostatic attraction between the two oppositely charged ions is the main force of an ionic bond. The atoms to form ionic bond must have different electronegativity large enough for the electron transfer. Examples include alkali metals (low electronegativity) and halogens (high electronegativity), LiF, **NaCl**, KI etc.

11. **(B)** In CO_2 molecule, there are two C=O double bonds. Carbon atom shares all its 4 valence electrons with two oxygen atoms, there is no lone pair of electron on carbon atom. Carbon atom has sp **hybridization**, and the CO_2 molecule is linear (**Figure** 4.1).

    The two oxygen atoms have two lone pairs of electron each besides the double bond with carbon atom, so oxygen atom has sp^2 **hybridization**.

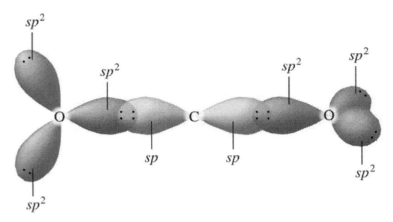

Figure 4.1 Hybridization in CO_2.

12. **(C)** In H_2O molecule, the central oxygen atom has sp^3 **hybridization**, two sp^3 orbitals form two covalent bonds with two hydrogen atoms, left two lone pairs of electrons. The two O–H bonds are at an angle. This geometry of H_2O molecule makes the molecule polar (Ḧ Ö H). All other choices are nonpolar.

13. **(E)** A few elements on periodic table show both metallic and nonmetallic properties, and have conductivity between nonmetals and metals. These elements are called semiconductor, including boron (B), **silicon** (Si), germanium (Ge), arsenic (As) etc.

14. **(D)** First ionization energy increase from left to right in a row, and decrease downward in a group. Therefore, **helium** (He) has the greatest first ionization energy (**Figure** 1.3) in all elements.

15. **(A)** Electronegativity follow similar trend with first ionization energy, but not include the noble gases (see above). **Fluorine** has the greatest electronegativity (**Figure** 2.2) in all elements.

16. **(C)** Transition metals form salt or complex which usually have color, because the transition metals has partially filled d orbitals, and d electrons can be easily excited to absorb visible light. Fe is the only transition metal in the list and is the correct answer.

17. **(B)** Only $AgNO_3$ reacts with NaCl solution to form white precipitate, AgCl, which is insoluble salt.

18. **(D) A coordinate bond** (also called a dative covalent bond) is a covalent bond in which both electrons come from the same atom. Nitrogen atom in ammonia, NH_3, forms three N–H bonds with a lone pair of electron. When NH_3 is dissolved in water, a proton can be attracted by the lone pair of electrons and form ammonium, NH_4^+. This forth N–H bond is different from the other three N–H bonds in that the two electrons shared between N and H are all from N atom. This N–H bond is called coordinate bond. Although, once this bond formed, it is virtually the same as other three N–H bond, or there is no way to distinguish it from three other N–H bonds. \

$$H-\overset{..}{\underset{H}{N}}-H \; + \; [\,H\,]^+ \longrightarrow \; H-\overset{\overset{H}{\uparrow}}{\underset{H}{N}}-H$$

19. **(E)** When HCl is dissolved in water, it forms a strong acid.

20. **(A)** $CaCO_3$ can react with diluted acid to release CO_2 gas which is colorless.

21. **(C)** KCl can be formed through neutralization reaction between a strong acid, HCl, and a strong base, KOH.

22. **(C)** Acetic acid is a weak acid, and only partially dissociated when dissolved in water. In solution, the dissociated acetate ion is **at equilibrium** with non-dissociated acetic acid as showed by:

23. **(A)** Combustion of magnesium releases heat, and therefore, is an **exothermic reaction**.

24. **(B)** A reaction which needs heat to be added is called **endothermic reaction**.

25. **(A)** When concentrated sulfuric acid is dissolved in water, heat is released, and this process is also **exothermic** process.

26. **(B)** Atom size (radius) decreases from left to right in a row (period). Electron affinity (C), ionization energy (D), number of valence electrons (E) all increase from left to right in a period. (A) is incorrect because the most reactive metals are on the left side, while most reactive nonmetal is the second column from right side of periodic table.

27. **(D)** For similar weak acid, the higher the K_a value, the more acid molecules dissociate, and the better electrolyte if the concentration is the same. However, both concentration and K_a are different in this problem. There is a way to compare without calculating the concentrations of dissociated ions.

    First, compare solution D with A, B and E. Since both concentration and K_a of solution D is greater than those of A, B and E. D must be better electrolyte than solutions A, B and E.

    Second, compare solution D with C. Although concentration of solution C is higher than that of D, acid C is so weak that the higher concentration of C will not help C to be better electrolyte than D.

28. **(C)** Given $[OH^-] = 10^{-8}$ M

    Since $[H^+][OH^-] = 10^{-14}$, $[H^+] = 10^{-14}/10^{-8} = 10^{-6}$ M.
    pH $= -\log[H^+] = -\log(10^{-6}) = $ **6**. This is an acidic solution.

29. **(D)** Barium phosphate is insoluble in water, and will form precipitate until either Ba^{2+} or PO_4^{3-} is completely consumed. The reaction equation is:

    $(NH_4)_3PO_4(aq) + Ba(NO_3)_2(aq) \rightarrow Ba_3(PO_4)_2(s) + NH_4NO_3(aq)$

30. **(A)** The diatomic molecule of Br_2 is nonpolar, and has no permanent polarity. The only force between Br_2 molecules is van der Waals force (dispersion force). Since the Br_2 molecules is relatively large (relative to Cl_2 and F_2), the temporary dipole is stronger, and so on the intermolecular van der Waals force.

31. **(C)** To completely titrate a strong acid with a strong base, the amount of OH^- added must equal to the amount of H^+ in the acid, i.e. $V_{acid} M_{acid} = V_{base} M_{base}$ where V_{acid} and M_{acid} are volume and molarity of acid, and V_{base} and M_{base} are volume and molarity of base used.

    $V_{base} = 21.05$ mL $- 6.05$ mL $= 15.00$ mL

    $M_{acid} = V_{base} M_{base}/ V_{acid} = 15$ mL x $(1.00$ M$)/(100$ mL$) = 0.15$ M.

32. **(E)** Alkanes are saturated hydrocarbon (C) which only contain C and H (A). They are nonpolar molecules (D), and are **insoluble in water** (E is incorrect). CH_4 is the simplest alkane (B).

33. **(E)** The process represented by the arrow is conversion of solid directly to gas, which is defined as sublimation (see **Figure** 1.1).

34. **(D)** According to the equation, there is 114 kJ heat is released for every 2 mol of NO_2 produced.

    For 0.5 mol NO_2 produced, heat released $= 0.5$ mol $NO_2 \times \dfrac{114 \text{ kJ}}{2 \text{ mol } NO_2} = 28.5$ kJ.

35. **(D)** The total mass of O in a $C_6H_{12}O_6$ molecule is $16 \times 6 = 96$. The percent mass of O in $C_6H_{12}O_6$ is $(96/180) \times 100\% = 53\%$.

36. **(A)** Electron affinity of an atom is defined as the amount of energy released when an electron is added to a neutral atom or molecule in the gaseous state to form a negative ion. The electron affinity increases from left to right in a row, and decreases downward in a group (this is the same with electronegativity, **Figure** 1.3). Therefore, F has the greatest affinity for electron.

37. **(A)** At STP, 1 mole of gas occupies 22.4 liter. 11.2 liter of gas is equivalent to 0.5 mol. Therefore, the molar mass of this gas is 22.0 grams/0.5 mol = 44.0 gram/mol. CO_2 is the only gas given which has molar mass of 44.0.

38. **(D)** The reactant Y and product W are solid, and should not be included in the equilibrium constant equation. The correct answer is D.

39. **(A)** The strength of carbon–carbon bond follows the order, $C\equiv C > C=C > C–C$. Ethyne (C_2H_2) has a triple bond, and has the strongest carbon–carbon bond.

40. **(C)** Molar mass of NaCl is $23.0 + 35.5 = 58.5$ (gram/mol). 58.5 g NaCl equals to 1 mol NaCl. Molarity of NaCl solution is 1 mol/1.5 L = 0.67 M.

41. **(E)** For this question, figure out number of atoms and add together to get the total.

Number of atoms represented by $Cr(NH_3)_5SO_4Br = 1 + (1 + 3) \times 5 + 1 + 4 + 1 = 27$.

42. **(C)** For ideal gas law problems, you need to get familiar with different variations of the general equation, $PV = nRT$. A typical type of question is about how one parameter changes after one or more other parameters (P, V, T, n) change. In this case, write two equations for both conditions, and plug in any known numbers and solve the equation.

	P	V	T (K)	n
Condition 1	450 mmHg	10 L	$273 + 50 = 323$	No change
Condition 2 (STP)	760 mmHg	?	273	No change

For this question, follow the steps below:

Step 1: Write gas law equations for both conditions: $P_1V_1 = nRT_1$ (1) and $P_2V_2 = nRT_2$ (2)

Step 2: Divide equation (1) by (2): $P_1V_1/P_2V_2 = T_1/T_2$ (3)

Step 3: Rearrange equation (3) to solve unknown:

$$V_2 = P_1V_1T_2/P_2T_1 = V_1 \times (P_1/P_2) \times (T_2/T_1) = 10 \times (450/760) \times (273/323)$$

43. **(B)** Concentration of 0.1 M and pH = 2.9 combined indicate solution II is a weak acid. If not, its pH is

$-\log[H^+] = -\log 0.1 = 1$, not 2.9.

pH of solution I indicates a strong base. Since the concentration (0.1 M) match pH (14), indicating the base dissociated completely.

pH of solution III indicates a strong acid. Since the concentration (0.01 M) match pH (2.0), indicating the acid dissociated completely.

44. **(D)** When the number of mol of gas in both sides of the equilibrium equation are the same, pressure would not influence the equilibrium. After balancing all the equilibrium equations, only choice D meet this condition.

(A) $\underline{3}\,O_2(g) \rightleftharpoons \underline{2}\,O_3(g)$ #mole decrease

(B) $N_2(g) + \underline{3}\,H_2(g) \rightleftharpoons \underline{2}\,NH_3(g)$ #mole decrease

(C) $\underline{2}\,NO_2(g) \rightleftharpoons N_2O_4(g)$ #mole decrease

(D) $H_2(g) + I_2(g) \rightleftharpoons \underline{2}\,HI(g)$ **#mole not changed**

(E) $\underline{2}\,CO(g) + O_2(g) \rightleftharpoons \underline{2}\,CO_2(g)$ #mole decrease

45. **(D)** The molar ratio of carbon to hydrogen in unknown hydrocarbon is calculated as

    C : H = (6/12)/(1/1) = 1/2. Only choice **D** (C_2H_4) has carbon to hydrogen ratio of 1:2.

    Approach above is a simplified one. Questions like this (known mass, mass ratio or mass percentage) involve calculation molar ratio of elements in a compound. A complete solution is showed below:

    Step 1: Assume that a sample of this compound has H with mass X, then the mass of C is 6X

    Step 2: Number of mol of H = X/(1 g/mol), number of mol of C = 6X/(12 g/mol)

    Step 3: Molar ratio C : H = $\dfrac{6X/(12\ g/mol)}{X/(1\ g/mol)} = \dfrac{1}{2}$

    Step 4: Determine the formula: C_2H_4 has molar ratio 1 : 2.

46. **(C)** The balanced equation is: $\underline{2}\,C_3H_8O + \underline{9}\,O_2 \rightarrow \underline{6}\,CO_2 + \underline{8}\,H_2O$

47. **(B)** $CaCl_2$ has i value of 3, greatest in all compounds given, therefore, $CaCl_2$ is most effective in melting ice. This question is about depression of freezing point, see **Table** 1.4 for details. In practice, NaCl is commonly used for melting snow due to its availability and the fact that NaCl is more environment friendly.

48. **(A)** Follow the steps below:

    Step 1: Number of mol of oxygen produced: (96 gram O_2)/(32 gram O_2/mol O_2) = 3 mol O_2.

    Step 2: According to the balanced equation, ratio of $C_6H_{12}O_6$ to O_2 produced in the reaction is 1:6. Number of moles of $C_6H_{12}O_6$ produced is 3 mol O_2 x (1 mol $C_6H_{12}O_6$ /6 mol O_2) = 0.5 mol.

    Step 3: Mass of $C_6H_{12}O_6$ produced:

    (0.5 mol $C_6H_{12}O_6$) x (180 grams $C_6H_{12}O_6$ /mol $C_6H_{12}O_6$) = 90 grams $C_6H_{12}O_6$.

49. **(C)** Catalyst only changes the activation energy of both the reactants and products, but **not** the **potential energies (III is correct)** of both reactants and products. Therefore, catalyst will not change the heat of the reaction which is the difference of the total potential energies between products and reactants. In addition, catalyst can **speed up** the process to **reach equilibrium** (II is incorrect), but will **not** change the **equilibrium constant (I is correct)**.

50. **(C)** According to the balanced equation, Pb and S react at 1:1 ratio. There is 20.7 grams of Pb, which is equivalent to (20.7 gram)/(207 gram/mol) = 0.1 mol; and 6.4 grams of S, which is equivalent to (6.4 gram)/(32 gram/mol) = 0.2 mol. Therefore, S is excessive and Pb is limiting reagent. After 0.1 mol Pb reacts completely, only 0.1 mol S is consumed, left 0.1 mol S unreacted. There will be (0.1 mol S) x (32 gram S/mol S) = 3.2 gram S left.

51. **(E)** The **kinetics molecular theory** by definition can be used for all states of matter, gas, liquid and solid. It states that all matters are made up of particles which are in constant motion. When kinetics molecular theory is used for gas, it explains some general properties of gas such as why gas can fill a container (diffusion and effusion) and how a gas exerts pressure on the wall of a container (kinetic collision).

52. **(E)** When using **Bunsen burner** to heat and dry a salt hydrate, a tong should be used to hold the crucible before and after the salt is dried. A wire gauze should be used to support the crucible. An analytical balance is used to measure mass before and after the salt hydrate is dried. Only thermometer is not necessary (it is might be useful if you want monitor the temperature in the crucible).

53. **(B)** Follow the step below:

Step 1: Balance the equation: $Fe_2O_3(s) + 3CO(g) \rightarrow 2Fe(s) + 3CO_2(g)$
Step 2: Number of mol of CO consumed: (56 gram CO)/(28 gram CO /mol CO) = 2 mol CO.
Step 3: Calculate number of mol produced: 2 mol CO x (2 mol Fe/3 mol CO) = 1.3 mol Fe.

54. **(A)** At STP, 4 grams H_2 equals to 2 mol H_2, and it occupies 22.4 x 2 = 44.8 liter of space. 4 grams He equals to 1 mol He, and occupies 22.4 liter of space. The calculation is summarized in following table:

Gas	Mass (gram)	Mole	Volume (liter)	Number of molecules
H_2	4	2	44.8	2 x 6.02 x 10^{23}
He	4	1	22.4	6.02 x 10^{23}

55. **(D)** Section D of the heating curve represents evaporation process in which liquid is converted to gas with temperature remains constant. Temperature will start to rise after all liquid is evaporated; before that, liquid and gas (vapor) exist at the same time.

56. **(B)** To balance redox reaction, not only masses of all elements need to be balanced, gain and loss of electrons also need to be balanced. In acidic or basic solution, H_2O may be needed to add to one side of the equation to balance H and O.

In neutral solution, no acid or base involved, a half equation method may be used to balance redox reaction (**Table** 4.1)

Table 4.1 *Balance Redox Reaction*
Balance redox reaction equation: $Ce^{3+} + Pb \rightarrow Ce + Pb^{4+}$
Step 1: Separate the half–reactions. $\quad Ce^{3+}(aq) + 3e^- \rightarrow Ce(s)$ (1) $\quad Pb(s) \rightarrow Pb^{4+}(aq) + 4e^-$ (2)
Step 2: Balance the electrons in the half equations. In this case, half equation (1) is multiplied by 4 and half equation (2) is multiplied by 3. Numbers of gain (3) or loss (4) of electrons are the same. $\quad 4Ce^{3+}(aq) + 12e^- \rightarrow 4Ce(s)$ (3) $\quad 3Pb(s) \rightarrow 3Pb^{4+}(aq) + 12e^-$ (4)
Step 3: Adding the half equations: $\quad 4Ce^{3+}(aq) + 12e^- + 3Pb(s) \rightarrow 4Ce(s) + 3Pb^{4+}(aq) + 12e^-$
Step 4: The electrons cancel out: $\quad 4Ce^{3+}(aq) + 3Pb(s) \rightarrow 4Ce(s) + 3Pb^{4+}(aq)$

57. **(D)** Molecular geometry, 3D configuration and central atom hybridization of CH_4, H_2O and $BeCl_2$ are summarized in table below:

Molecule	Geometry	3D configuration	Hybridization
CH_4	Tetrahedral		sp^3
H_2O	Bent		sp^4
$BeCl_2$	Linear	$\ddot{C}l-Be-\ddot{C}l$	sp

58. **(D)** An endothermic reaction is any chemical reaction that absorbs heat from environment. It needs to absorb energy to proceed since the total potential energy of products is higher than that of reactants, the energy absorbed is converted to excessive potential energy of products. This is reflected on the reaction diagram in which the energy level of products is higher than that of reactants. Therefore, it is the difference between A and C on the diagram provides information on whether it's exothermic and endothermic. Potential energy of either reactants or products alone cannot tell if it's an exothermic or endothermic reaction. The difference between A and B, and between C and B provide information on activation energy for forward and reverse reactions, but don't tell if this reaction is exothermic or endothermic.

59. **(C)** is correctly named.

(A) CO — ~~mono~~carbon monoxide, INCORRECT. → carbon monoxide
(B) CaF_2 — calcium ~~di~~fluoride, INCORRECT. → calcium fluoride
(C) CCl_4 — carbon tetrachloride, CORRECT
(D) PCl_3 — potassium ~~tri~~chloride, INCORRECT. → phosphorus chloride
(E) TiF_4 — tin(IV) ~~tetra~~fluoride, INCORRECT. → titanium(IV) fluoride (Ti is titanium, not tin).

60. **(D)** O_2 is only choice which does not react with $NaOH$.

61. **(B)** All hydrocarbon are neutral molecules, there are no permanent dipole–dipole interaction between them. **van der Waals force** (dispersion force) increases with size of molecules, and attraction due to temporary dipole increases with size. That's why small hydrocarbon has lower melting point.

62. **(D)** NH_3 is the only choice given which is readily dissolved in water (Cl_2 is another common gas which is not collected via water displacement). CO_2 is also soluble in water, but its solubility is much lower than NH_3. All other gases (CH_4, H_2, O_2) are considered insoluble in water.

63. **(A)** Robert Millikan is the first scientist who observed charge of **electrons** through the famous oil drop experiment.

64. **(E)** All three process increase randomness of the system, and cause increase in entropy.

I. When dissolve salt in water, two substances mix together and increase randomness.
II. Sublimation is process when a solid convert to gas directly. Gas has greater randomness than liquid and solid.
III. Heating up a liquid increase kinetic energy of molecules in liquid, and increase randomness.

65. **(B)** Either increasing the pressure or decreasing the temperature will cause the volume of a gas to decrease based on ideal gas law, $V = nRT/P$. If both processes apply at the same time, it will always decrease the volume of the gas (II is correct). When decreasing the pressure while holding temperature constant, the volume will increase (option I is incorrect). If both pressure and temperature increase, the effect is a mixture, since these two processes have opposite effect (III is incorrect), and the volume might increases, decreases, or remain the same.

66. **(B)** This question is about stoichiometry of heat, follow steps below:

Step 1: Calculate number of mol of O_2: $32 \text{ g } O_2 \times \frac{1 \text{ mol } O_2}{32 \text{ g } O_2} = 1 \text{ mol } O_2$.

Step 2: Calculate heat released: $1 \text{ mol } O_2 \times \frac{800 \text{ kJ}}{2 \text{ mol } O_2} = 400 \text{ kJ}$.

Combined calculation: $32 \text{ g } O_2 \times \frac{1 \text{ mol } O_2}{32 \text{ g } O_2} \times \frac{800 \text{ kJ}}{2 \text{ mol } O_2} = 400 \text{ kJ}$

67. **(D)** Since the other end of the manometer is open, the pressure of gas in the enclosed container equals to the sum of atmospheric pressure and the difference in the height of the mercury column:

Pressure of gas: 760 mmHg + 180 mmHg = 940 mmHg.

68. **(D)** Under the same pressure and temperature condition, the density of a gas is determined by its molar mass; therefore, H_2 has the lowest density. In addition, the rate of effusion decrease with increase in the size of the molecule, and H_2 has the highest rate effusion.

69. **(B)** According to quantum mechanics, each electron in an atom is associated with a definite energy level (called orbitals). The energy levels are described by a series number called quantum numbers, which describe the size, shape, and orientation in space of the orbitals on an atom. There are no definite orbits for electron to occupy, and electron is keeping moving without fixing at any given position.

70. **(B)** For multiple step reaction, the slowest step makes the overall reaction slow and is the limiting step. The slow reaction rate is caused by high activation energy. The higher the activation energy, the less reactants molecules are activated for the reaction to occur, and the lower the reaction rate.

101. **Correct answer: I True, II True, CE Yes**

    Explanation: In Rutherford gold foil experiment, a beam of alpha particles directed onto a sheet of very thin gold foil in an evacuated chamber. It's found that small fraction of alpha particles were deflected and even reflected. This was used as evidence that the mass of an atom (gold) concentrated in a very small central particles (nucleus). If the alpha particle collide with the nucleus head–on–head, it will be repelled (reflected). This is one of a series famous experiments which led to the modern atomic theory.

102. **Correct answer: I False, II True, CE No**

    Explanation: The atomic number of nitrogen is 7 with 5 valance electrons (Statement II is true). The correct electron configuration of N is $1s^2 2s^2 2p^3$.

103. **Correct answer: I False, II False, CE No**

    Explanation: The hybridization in H_2O is sp^3 (Statement II is false), 2 electrons form covalence bonds with 2 hydrogen atoms, leaving two lone pairs of electron. That's why H_2O has bent geometry rather than linear geometry (statement I is incorrect).

104. **Correct answer: I False, II False, CE No**

    Explanation: AgCl is insoluble in water (statement I is false), although most chloride (not all) salts are soluble (statement II is false).

105. **Correct answer: I False, II True, CE No**

    Explanation: Although C–Cl bond in CCl_4 is polar (statement II is true), CCl_4 is nonpolar (statement I is false) due to the geometry (tetrahedral).

106. **Correct answer: I True, II True, CE Yes**

    Explanation: In solid KCl crystal, both K^+ and Cl^- ions are fixed, and cannot move freely. Once melt, K^+ and Cl^- is relatively mobile, and can conduct electricity. Statement II is appropriate explanation of statement I.

107. **Correct answer: I True, II True, CE Yes**

    Explanation: When HCl dissolved in water, it releases hydronium ion H_3O^+. This fits the definition of Arrhennius acid. Statement II is appropriate explanation of statement I.

108. **Correct answer: I True, II False, CE No**

    Explanation: Adding more reactants will **increase** the frequency that reactants collide with each other, and speed up the reaction. Therefore, statement I is true, and statement II is false.

109. **Correct answer: I False, II True, CE No**

Explanation: The reaction for Cu^{2+} to gain 2 electrons to become Cu is a **reduction reaction** rather than an oxidation reaction (Statement I is false). However, the equation is properly balanced (Statement II is true).

110. **Correct answer: I True, II False, CE No**

   Explanation: The decimals in 15.00 mL indicates precision of the measurement, in this case, two decimal indicates that the measurement can be precise to the 1/100 level, which reflects the precision of the buret reading (Statement I is true). However, decimal of reading depends on precision of device used, and not all readings needed to be precise to 1/100 level (Statement II is false).

111. **Correct answer: I True, II True, CE Yes**

   Explanation: Electronegativity is the tendency to attract electron, it increases from left to right in a row and decreases down a column. Therefore, F has the highest electronegativity, and F has the greatest attraction for electron. Statement II is appropriate explanation of statement I.

112. **Correct answer: I False, II True, CE No**

   Explanation: When sugar is dissolved in water, the force between molecules (van der Waals forces, or dipole–dipole forces etc.) is broken, but the covalent bonds in the molecules are not affected (Statement I is false). Statement II is the definition of covalent bond, and it's true.

113. **Correct answer: I False, II True, CE No**

   Explanation: The radius increases from left to right in a row, therefore, sodium has larger radius than chlorine (Statement I is false). Sodium has only one valence electron, while chlorine has 7 valence electron (Statement II is true).

114. **Correct answer: I True, II True, CE Yes**

   Explanation: Sodium has one valence electron, and readily loses this electron to be oxidized, therefore, sodium is a good reducing reagent. Both statement I and II are true, and statement II is appropriate explanation of statement I.

115. **Correct answer: I True, II False, CE No**

   Explanation: Salt can depress melting/freezing point, and solution of salt freeze below 0°C (Statement I is true). Salt may also elevate boiling point, not lower boiling point (Statement II is false).

SAT II Chemistry

Practice Test 5

Answer Sheet

Part A and C: Determine the correct answer. Blacken the oval of your choice completely with a No. 2 pencil.

1	Ⓐ Ⓑ Ⓒ Ⓓ Ⓔ	25	Ⓐ Ⓑ Ⓒ Ⓓ Ⓔ	49	Ⓐ Ⓑ Ⓒ Ⓓ Ⓔ
2	Ⓐ Ⓑ Ⓒ Ⓓ Ⓔ	26	Ⓐ Ⓑ Ⓒ Ⓓ Ⓔ	50	Ⓐ Ⓑ Ⓒ Ⓓ Ⓔ
3	Ⓐ Ⓑ Ⓒ Ⓓ Ⓔ	27	Ⓐ Ⓑ Ⓒ Ⓓ Ⓔ	51	Ⓐ Ⓑ Ⓒ Ⓓ Ⓔ
4	Ⓐ Ⓑ Ⓒ Ⓓ Ⓔ	28	Ⓐ Ⓑ Ⓒ Ⓓ Ⓔ	52	Ⓐ Ⓑ Ⓒ Ⓓ Ⓔ
5	Ⓐ Ⓑ Ⓒ Ⓓ Ⓔ	29	Ⓐ Ⓑ Ⓒ Ⓓ Ⓔ	53	Ⓐ Ⓑ Ⓒ Ⓓ Ⓔ
6	Ⓐ Ⓑ Ⓒ Ⓓ Ⓔ	30	Ⓐ Ⓑ Ⓒ Ⓓ Ⓔ	54	Ⓐ Ⓑ Ⓒ Ⓓ Ⓔ
7	Ⓐ Ⓑ Ⓒ Ⓓ Ⓔ	31	Ⓐ Ⓑ Ⓒ Ⓓ Ⓔ	55	Ⓐ Ⓑ Ⓒ Ⓓ Ⓔ
8	Ⓐ Ⓑ Ⓒ Ⓓ Ⓔ	32	Ⓐ Ⓑ Ⓒ Ⓓ Ⓔ	56	Ⓐ Ⓑ Ⓒ Ⓓ Ⓔ
9	Ⓐ Ⓑ Ⓒ Ⓓ Ⓔ	33	Ⓐ Ⓑ Ⓒ Ⓓ Ⓔ	57	Ⓐ Ⓑ Ⓒ Ⓓ Ⓔ
10	Ⓐ Ⓑ Ⓒ Ⓓ Ⓔ	34	Ⓐ Ⓑ Ⓒ Ⓓ Ⓔ	58	Ⓐ Ⓑ Ⓒ Ⓓ Ⓔ
11	Ⓐ Ⓑ Ⓒ Ⓓ Ⓔ	35	Ⓐ Ⓑ Ⓒ Ⓓ Ⓔ	59	Ⓐ Ⓑ Ⓒ Ⓓ Ⓔ
12	Ⓐ Ⓑ Ⓒ Ⓓ Ⓔ	36	Ⓐ Ⓑ Ⓒ Ⓓ Ⓔ	60	Ⓐ Ⓑ Ⓒ Ⓓ Ⓔ
13	Ⓐ Ⓑ Ⓒ Ⓓ Ⓔ	37	Ⓐ Ⓑ Ⓒ Ⓓ Ⓔ	61	Ⓐ Ⓑ Ⓒ Ⓓ Ⓔ
14	Ⓐ Ⓑ Ⓒ Ⓓ Ⓔ	38	Ⓐ Ⓑ Ⓒ Ⓓ Ⓔ	62	Ⓐ Ⓑ Ⓒ Ⓓ Ⓔ
15	Ⓐ Ⓑ Ⓒ Ⓓ Ⓔ	39	Ⓐ Ⓑ Ⓒ Ⓓ Ⓔ	63	Ⓐ Ⓑ Ⓒ Ⓓ Ⓔ
16	Ⓐ Ⓑ Ⓒ Ⓓ Ⓔ	40	Ⓐ Ⓑ Ⓒ Ⓓ Ⓔ	64	Ⓐ Ⓑ Ⓒ Ⓓ Ⓔ
17	Ⓐ Ⓑ Ⓒ Ⓓ Ⓔ	41	Ⓐ Ⓑ Ⓒ Ⓓ Ⓔ	65	Ⓐ Ⓑ Ⓒ Ⓓ Ⓔ
18	Ⓐ Ⓑ Ⓒ Ⓓ Ⓔ	42	Ⓐ Ⓑ Ⓒ Ⓓ Ⓔ	66	Ⓐ Ⓑ Ⓒ Ⓓ Ⓔ
19	Ⓐ Ⓑ Ⓒ Ⓓ Ⓔ	43	Ⓐ Ⓑ Ⓒ Ⓓ Ⓔ	67	Ⓐ Ⓑ Ⓒ Ⓓ Ⓔ
20	Ⓐ Ⓑ Ⓒ Ⓓ Ⓔ	44	Ⓐ Ⓑ Ⓒ Ⓓ Ⓔ	68	Ⓐ Ⓑ Ⓒ Ⓓ Ⓔ
21	Ⓐ Ⓑ Ⓒ Ⓓ Ⓔ	45	Ⓐ Ⓑ Ⓒ Ⓓ Ⓔ	69	Ⓐ Ⓑ Ⓒ Ⓓ Ⓔ
22	Ⓐ Ⓑ Ⓒ Ⓓ Ⓔ	46	Ⓐ Ⓑ Ⓒ Ⓓ Ⓔ	70	Ⓐ Ⓑ Ⓒ Ⓓ Ⓔ
23	Ⓐ Ⓑ Ⓒ Ⓓ Ⓔ	47	Ⓐ Ⓑ Ⓒ Ⓓ Ⓔ	71	Ⓐ Ⓑ Ⓒ Ⓓ Ⓔ
24	Ⓐ Ⓑ Ⓒ Ⓓ Ⓔ	48	Ⓐ Ⓑ Ⓒ Ⓓ Ⓔ	72	Ⓐ Ⓑ Ⓒ Ⓓ Ⓔ

Part B: On the actual Chemistry Test, the following type of question must be answered on a special section (labeled "Chemistry") at the lower left–hand corner of your answer sheet. These questions will be numbered beginning with 101 and must be answered according to the directions.

	I		II		CE
	PART B				
	I		II		CE
101	T	F	T	F	◯
102	T	F	T	F	◯
103	T	F	T	F	◯
104	T	F	T	F	◯
105	T	F	T	F	◯
106	T	F	T	F	◯
107	T	F	T	F	◯
108	T	F	T	F	◯
109	T	F	T	F	◯
110	T	F	T	F	◯
111	T	F	T	F	◯
112	T	F	T	F	◯
113	T	F	T	F	◯
114	T	F	T	F	◯
115	T	F	T	F	◯
116	T	F	T	F	◯

Periodic Table of Elements

Material in this table may be useful in answering the questions in this examination

1 H 1.0079																	2 He 4.0026
3 Li 6.941	4 Be 9.012											5 B 10.811	6 C 12.011	7 N 14.007	8 O 16.00	9 F 19.00	10 Ne 20.179
11 Na 22.99	12 Mg 24.30											13 Al 26.98	14 Si 28.09	15 P 30.974	16 S 32.06	17 Cl 35.453	18 Ar 39.948
19 K 39.01	20 Ca 40.48	21 Sc 44.96	22 Ti 47.90	23 V 50.94	24 Cr 52.00	25 Mn 54.938	26 Fe 55.85	27 Co 58.93	28 Ni 58.69	29 Cu 63.55	30 Zn 65.39	31 Ga 69.72	32 Ge 72.59	33 As 74.92	34 Se 78.96	35 Br 79.90	36 Kr 83.80
37 Rb 85.47	38 Sr 87.62	39 Y 88.91	40 Zr 91.22	41 Nb 92.91	42 Mo 95.94	43 Tc (98)	44 Ru 101.1	45 Rh 102.91	46 Pd 106.42	47 Ag 107.87	48 Cd 112.41	49 In 114.82	50 Sn 118.71	51 Sb 121.75	52 Te 127.60	53 I 126.91	54 Xe 131.29
55 Cs 132.91	56 Ba 137.33	57 *La 138.91	72 Hf 178.49	73 Ta 180.95	74 W 183.85	75 Re 186.21	76 Os 190.2	77 Ir 192.2	78 Pt 195.08	79 Au 196.97	80 Hg 200.59	81 Tl 204.38	82 Pb 207.2	83 Bi 208.98	84 Po (209)	85 At (210)	86 Rn (222)
87 Fr (223)	88 Ra 226.02	89 Ac 227.03	104 Rf (261)	105 Db (262)	106 Sg (266)	107 Bh (264)	108 Hs (277)	109 Mt (268)	110 Ds (271)	111 Rg (272)	112 (277)						

*Lanthanide Series

58 Ce 140.12	59 Pr 140.91	60 Nd 144.24	61 Pm (145)	62 Sm 150.4	63 Eu 151.97	64 Gd 157.25	65 Tb 158.93	66 Dy 162.50	67 Ho 164.93	68 Er 167.26	69 Tm 168.93	70 Yb 173.04	71 Lu 174.97

Actinide Series

90 Th 232.04	91 Pa 231.04	92 U 238.03	93 Np 237.05	94 Pu (244)	95 Am (243)	96 Cm (247)	97 Bk (247)	98 Cf (251)	99 Es (252)	100 Fm (257)	101 Md (258)	102 No (259)	103 Lr (260)

Note: For all questions involving solutions, assume that the solvent is water unless otherwise stated.

Reminder: You may not use a calculator in this test!

Throughout the test the following symbols have the definitions specified unless otherwise noted.

H = enthalpy	atm = atmosphere(s)
M = molar	g = gram(s)
n = number of moles	J = joule(s)
P = pressure	kJ = kilojoule(s)
R = molar gas constant	L = liter(s)
S = entropy	mL = milliliter(s)
T = temperature	mm = millimeter(s)
V = volume	mol = mole(s)
	V = volt(s)

Chemistry Subject Practice Test 5

Part A

Directions for Classification Questions
Each set of lettered choices below refers to the numbered statements or questions immediately following it. Select the one lettered choice that best fits each statement or answers each question and then fill in the corresponding circle on the answer sheet. A choice may be used once, more than once, or not at all in each set.

Questions 1 – 4 refer to the following substances

(A) Ionic substance
(B) Nonpolar covalent substance
(C) Polar covalent substance
(D) Macromolecular substance
(E) Metallic substance

1. Boron trifluoride
2. Strontium
3. Potassium iodide
4. Formaldehyde, HCHO

Questions 5 – 8 refer to the following organic structural formula

(A) R–COOH
(B) R–CHO
(C) R–CO–R'
(D) R–COO–R'
(E) R–OH

5. Can be neutralized with a base
6. Could be named 2–pentanone
7. Also called alcohol
8. Aldehyde functional group

Questions 9 – 11 refer to the following quantities

(A) 6.02×10^{23} molecules
(B) 11.2 liters
(C) 58.5 grams/mole
(D) 2.0 moles
(E) 5 atoms

9. 88 grams of $CO_2(g)$ at STP
10. 1.0 molecule of CH_4
11. 32 grams of SO_2 gas at STP

Questions 12 – 15 refer to the following groups of periodic table

(A) Alkali metal
(B) Alkaline earth metal
(C) Transition metal
(D) Halogen
(E) Noble gas

12. Reacts most vigorously with water
13. Has greatest electronegativity value in its period
14. Has the highest first ionization energy in its period
15. Form oxides with formula X_2O

GO ON TO THE NEXT PAGE

Questions 16 – 18 refer to the following colors

    (A) Blue
    (B) Red
    (C) Pink/purple
    (D) Colorless
    (E) Orange

16. phenolphthalein in base

17. litmus in acid

18. phenolphthalein in acid

Questions 19 – 22 refer to the following gas laws

    (A) Boyle's law
    (B) Charle's law
    (C) Ideal gas law
    (D) Combined gas law
    (E) Dalton's law of partial pressures

19. $P_{total} = P_1 + P_2 + P_3 + \ldots$

20. $P_1V_1 = P_2V_2$

21. $PV = nRT$

22. $P_1V_1/T_1 = P_2V_2/T_2$

Questions 23 – 25 refer to the following signs

    (A) ^4_2He

    (B) $^0_{-1}e$

    (C) $^0_0\gamma$

    (D) ^1_0n

    (E) ^1_1H

23. emitted when U–238 decay to Th–234

24. is a high energy photon

25. emitted when C–14 decay to N–14

GO ON TO THE NEXT PAGE

Practice Test 5———*Continued*

PLEASE GO TO THE SPECIAL SECTION AT THE LOWER LEFT–HAND CORNER OF PAGE 2 OF YOUR ANSWER SHEET LABELED CHEMISTRY AND ANSWER QUESTIONS 101–115 ACCORDING TO THE FOLLOWING DIRECTIONS.

Part B

Directions for Relationship Analysis Questions
Each question below consists of two statements, I in the left–hand column and II in the right–hand column. For each question, determine whether statement I is true or false and whether statement II is true or false and fill in the corresponding T or F circles on your answer sheet. *Fill in circle CE only if statement II is a correct explanation of the true statement I.

EXAMPLES:

I		II
EX1. The nucleus in an atom has a positive charge.	BECAUSE	Proton has positive charge, neutron has no charge.

SAMPLE ANSWERS

		I		II		CE
EX1		●	Ⓕ	●	Ⓕ	●

	I		II
101.	^{14}C is an isotope of ^{14}N	BECAUSE	the nuclei of both atoms have the same mass number.
102.	Ice is less dense than liquid water	BECAUSE	water molecules are polar.
103.	A solution with a pH of 5 is less acidic than a solution with a pH of 8	BECAUSE	a solution with a pH of 5 has 1000 times less H_3O^+ ions than a solution with a pH of 8.
104.	Transition metal compounds are often colored	BECAUSE	transition metals often possess partially filled d orbitals.
105.	A voltaic cell spontaneously converts chemical energy into electrical energy	BECAUSE	in a voltaic cell, oxidation and reduction reaction occur in separate containers.
106.	The F^- ion has similar chemical properties with neon atom	BECAUSE	The F^- ion and the neon atom have the same number of electrons.

145

107. At equilibrium the concentration of reactants and products remain unchanged   BECAUSE   at equilibrium the rates of the forward and reverse reactions are all zero.

108. Elements in the same group have similar chemical properties   BECAUSE   the valence shells of elements in the same group have the same energy.

109. *n*–butane and 2–methylpropane are called isomers   BECAUSE   *n*–butane and 2–methylpropane both have the molecular formula of C_4H_{10}.

110. Fluorine is a stronger oxidizing reagent than chlorine   BECAUSE   fluorine atoms are smaller than chlorine atoms.

111. Ionic bond is formed between Cl^- ion and K^+ ion   BECAUSE   both Cl^- ion and K^+ ion has the same outmost electron configuration as argon atom.

112. Entropy increases when ice melts   BECAUSE   water molecules are distributed more randomly in liquid state than in sold state.

113. Ammonia has a trigonal pyramidal molecular geometry   BECAUSE   ammonia has a lone pair of electrons that repel bonding electrons.

114. $AlCl_3$ is called aluminum trichloride   BECAUSE   prefixes indicating number of atoms shown in formula are used when naming covalent compounds.

115. The combustion of fossil fuel containing sulfur leads to the production of acid rain   BECAUSE   sulfur oxide, SO_2, dissolves in water to produce an acidic solution.

GO ON TO THE NEXT PAGE

Part C

Directions for Five–Choice Completion Questions
Each of the questions or incomplete statements below is followed by five suggested answers or completions. Select the one that is best in each case and then fill in the corresponding circle on the answer sheet.

26. Complete combustion of hydrocarbon produces

    (A) CH_3OH
    (B) H_2O and CH_4
    (C) H_2 and CO_2
    (D) H_2O and CO_2
    (E) CH_3COOH

27. $MgCO_3(s) + HCl \rightarrow$

    Products of the reaction represented above include which of the following?

    I. $Mg(s)$
    II. $Mg^{2+}(aq)$
    III. $CO_2(g)$

    (A) I only
    (B) II only
    (C) III only
    (D) I and II only
    (E) II and III only

28. Which of the following best account for some of the non–ideal behavior observed in real gas?

    (A) Some gaseous molecules are not spherical.
    (B) There are intermolecular attractive forces.
    (C) The temperature is not kept constant.
    (D) R, the gas constant, is not a true constant.
    (E) Experimental errors are made in the measurement of the pressure and the volume.

29. The following redox reaction occurs in an acidic solution: $Ce^{4+} + Bi \rightarrow Ce^{3+} + BiO^+$. What is the coefficient before the Ce^{4+} when the equation is fully balanced and all coefficients are reduced to simplest whole number?

    (A) 1
    (B) 2
    (C) 3
    (D) 6

30. What is the density of pure oxygen gas at STP? ($R = 0.0821$ atm L $mol^{-1}K^{-1}$)

    (A) 0.14 g/L
    (B) 0.43 g/L
    (C) 0.85 g/L
    (D) 1.43 g/L
    (E) 4.20 g/L

31. Which statement below best describes the molecule in question?

    (A) Water has a bent molecular geometry and one lone pair of electrons.
    (B) Ammonia has a trigonal pyramidal molecular geometry and two lone pairs of electrons.
    (C) Methane has a trigonal planar molecular geometry.
    (D) Carbon dioxide is linear because it has one single bond and one triple bond.
    (E) The carbon atoms in ethane are sp^3 hybridized.

GO ON TO THE NEXT PAGE

32. Elemental composition of a compound by mass was analyzed as 12.1% carbon, 71.7% chlorine, and 16.2% oxygen. What is the empirical formula for this compound?

    (A) C_2OCl
    (B) $COCl$
    (C) CO_2Cl_2
    (D) C_2O_2Cl
    (E) CCl_2O

33. Which of the following represents the ground–state electron configuration of a neutral phosphorus (P) atom?

    (A) $1s^22s^22p^3$
    (B) $1s^22s^22p^63s^23p^1$
    (C) $1s^22s^22p^63s^23p^3$
    (D) $1s^22s^22p^63s^23p^64s^2$
    (E) $1s^22s^22p^63s^23p^63d^14s^2$

34. $H_2(g) + I_2(g) + 51.9 \text{ kilojoules} \rightarrow 2HI\ (g)$

    Which of the following can be expected to increase the rate of the reaction given by the equation above?

    I.   Adding some helium gas
    II. Adding a catalyst
    III. Increasing the temperature

    (A) I only
    (B) II only
    (C) III only
    (D) II and III only
    (E) I, II, and III

35. Which of the following statements about the halogens is true?

    (A) They all form X^- ions.
    (B) They have the lowest ionization energy of the elements in the respective period.
    (C) They are all gases at room temperature.
    (D) They are among the best reducing reagents.
    (E) They all react with water to form basic solution.

36. Which noble gas is expected to show the largest deviations from the ideal gas behavior?

    (A) Helium
    (B) Neon
    (C) Argon
    (D) Krypton
    (E) Xenon

37. Two aqueous solutions are prepared: 1.0 m $Cu(NO_3)_2$ and 1.0 m NaBr. Which of the following statements is true?

    A) The $Cu(NO_3)_2$ solution has a higher boiling point and lower freezing point than the NaBr solution.

    B) The $Cu(NO_3)_2$ solution has a higher boiling point and higher freezing point than the NaBr solution.

    C) The $Cu(NO_3)_2$ solution has a lower boiling point and lower freezing point than the NaBr solution.

    D) The $Cu(NO_3)_2$ solution has a lower boiling point and higher freezing point than the NaBr solution.

    E) None of the above statements is true.

38. The following table lists values for three properties of water. How much heat must be added to 10 grams of ice at 0°C to convert it to liquid water at 100°C?

Property	Value
Heat of fusion (cal/g)	80
Heat of vaporization (cal/g)	540
Specific heat (cal/g·°C)	1.00

    (A) 800 cal
    (B) 1,000 cal
    (C) 1,800 cal
    (D) 6,200 cal
    (E) 7,200 cal

GO ON TO THE NEXT PAGE

39. The following substances were all dissolved in 100 grams of water at 290 K to produce saturated solutions. If the solution is heated to 310 K, which substance will have a decrease in its solubility.

(A) NaCl
(B) KI
(C) $CaCl_2$
(D) HCl
(E) KNO_3

40. Methane undergoes a combustion reaction according to the reaction

$$CH_4(g) + 2O_2(g) \rightarrow CO_2(g) + 2H_2O(l).$$

How many grams of methane gas were burned if 67.2 liters of carbon dioxide gas are produced in the reaction? (assume STP)

(A) 3 grams
(B) 18 grams
(C) 54 grams
(D) 108 grams
(E) 162 grams

41. Increasing the temperature of an endothermic reaction results in _____

(A) more products and less reactants.
(B) more reactants and less products.
(C) more reactants and products.
(D) less reactants and products.
(E) no change in the quantities of reactants and products.

42. $Cl_2(g) + 2NO_2(g) \rightleftharpoons 2NO_2Cl(g)$.

At a particular temperature, the equilibrium concentrations of the substances in the previous question are as follows: $[NO_2Cl] = 0.5$ M, $[Cl_2] = 0.3$ M, $[NO_2] = 0.2$ M. What is the value of the equilibrium constant for this reaction?

(A) 2.1
(B) 0.48
(C) 0.036
(D) 20.8
(E) 208.3

43. With the Lewis structure of an ammonia molecules

H
H : N : H

Which of the following describes the shape of an ammonia molecule?

(A) T–shape
(B) Tetrahedral
(C) Square planar
(D) Trigonal planar
(E) Trigonal pyramidal

44. A 0.001 M solution of which of the following has a hydroxide ion (OH^-) concentration of 1×10^{-3}?

(A) $HC_2H_3O_2$
(B) HCl
(C) H_2SO_3
(D) NaOH
(E) KCl

45. The mass of 6.02×10^{23} molecules of a gas is 64.0 grams. What volume does 8.00 grams of the gas occupy at standard temperature and pressure?

(A) 2.80 liters
(B) 5.60 liters
(C) 11.2 liters
(D) 22.4 liters
(E) 64.0 liters

GO ON TO THE NEXT PAGE

149

46. Which pair of compounds below are isomers?

    (A) $CH_3CH_2CH_2OH$ and $HOCH_2CH_2CH_3$
    (B) $CH_3CH_2CH_3$ and $CH_3CH_2CH_2CH_3$
    (C) $CH_3CH(Cl)CH_3$ and $CH_3CH_2CH_2Cl$
    (D) CH_3COCH_3 and $CH_3CH_2CH_2CHO$
    (E) $ClCH_2CH_2Br$ and $BrCH_2CH_2Cl$

47. An atom contains 18 protons, 18 electrons, and 19 neutrons. Which of the following combinations of particles is an isotope of that atom?

	Proton	Electron	Neutron
(A)	18	19	19
(B)	18	18	18
(C)	17	17	17
(D)	19	18	19
(E)	19	19	18

48. $C_2H_6(g) + O_2(g) \rightarrow CO_2(g) + H_2O(g)$

    When the reaction above is completely balanced using the lowest whole number coefficients, what is the coefficient for O_2?

    (A) 4
    (B) 5
    (C) 6
    (D) 7
    (E) 8

49. To test if a water solution of $CuSO_4$ is supersaturated at room temperature, one can

    (A) Heat the solution to boiling point
    (B) Add water to the solution
    (C) Acidify the solution
    (D) Cool the solution to freezing point
    (E) Add a crystal of $CuSO_4$ to the solution

50. Which solutions have a concentration of approximately 1.0 M?

    I.   74.0 grams of calcium hydroxide dissolved to make 1 liter of solution
    II. 74.5 grams of potassium chloride dissolved to make 1 liter of solution
    III. 87 grams of lithium bromide dissolved to make 1 liter of solution

    (A) I only
    (B) III only
    (C) I and III only
    (D) II and III only
    (E) I, II and III

51. According to the reaction $3H_2(g) + N_2(g) \rightarrow 2NH_3(g)$, how many grams of hydrogen gas and nitrogen gas are needed to make exactly 68 grams of ammonia?

    (A) 2 grams of $H_2(g)$ and 28 grams of $N_2(g)$
    (B) 3 grams of $H_2(g)$ and 1 gram of $N_2(g)$
    (C) 12 grams of $H_2(g)$ and 56 grams of $N_2(g)$
    (D) 102 grams of $H_2(g)$ and 34 grams of $N_2(g)$
    (E) 6 grams of $H_2(g)$ and 2 grams of $N_2(g)$

52. A molecule of which of the following compounds contains a double bond?

    (A) C_2H_2
    (B) C_2H_4
    (C) C_3H_8
    (D) C_4H_{10}
    (E) C_2H_6O

GO ON TO THE NEXT PAGE

150

53. Which of the following pairs of acid/base produce KNO_3 after reacting in aqueous solution?

(A) $HClO_4$ + KOH
(B) HNO_3 + NaOH
(C) HNO_3 + KOH
(D) KCl + HNO_3
(E) HCl + NaOH

54. 200 mL water is added to 400 mL of a 0.6 M solution. What is the molarity of this diluted solution?

(A) 0.50 M
(B) 0.40 M
(C) 0.30 M
(D) 0.13 M
(E) 0.10 M

55. Which mixture is correctly paired with a method for separation of the mixture?

(A) Oil and water—filter paper
(B) Salt water—distillation
(C) Sand and water—separatory funnel
(D) Sand and sugar— filter paper
(E) Sugar water—centrifuge

56. Which reaction between ions does not form a precipitate?

(A) Ag^+ and Cl^-
(B) Pb^{2+} and PO_4^{3-}
(C) Ca^{2+} and CO_3^{2-}
(D) Hg^{2+} and Br^-
(E) Rb^+ and I^-

57. Which will happen when sodium sulfate is added to a saturated solution of $CaSO_4$ that is at equilibrium?

$$CaSO_4(s) \rightleftharpoons Ca^{2+}(aq) + SO_4^{2-}(aq)$$

(A) The solubility of the calcium sulfate will increase
(B) The concentration of calcium ions will increase
(C) The reaction will shift to the right
(D) The K_{sp} value will change
(E) The equilibrium will shift to the left.

58. Iron (III) chloride hexahydrate is used as a coagulant for sewage and industrial wastes. What is its formula?

(A) $Fe(Cl \cdot 6H_2O)_3$
(B) $Fe_3Cl \cdot 6H_2O$
(C) $FeCl_3(H_2O)_6$
(D) $Fe_3Cl(H_2O)_6$
(E) $FeCl_3 \cdot 6H_2O$

59. Which of the following gases can properly be dried by means of NaOH?

(A) CO_2
(B) H_2S
(C) HBr
(D) CH_4
(E) HCl

60. Which of these substances can sublime?

I.   Iodine
II. Naphthalene
III. Carbon dioxide

(A) I only
(B) II only
(C) III only
(D) I and III only
(E) I, II and III

GO ON TO THE NEXT PAGE

151

61. Given 5 reactant mixture

Mixture	mL of 0.2 M $Pb(NO_3)_2$	mL of 0.2 M NaI
I	2	7
II	3	6
III	4	5
IV	5	4
V	6	3

Which mixture yields the maximum amount of precipitated PbI_2?

(A) I (B) II (C) III (D) IV (E) V

62. A sealed container has 0.3 moles of gas A, 0.4 moles of gas B, and 0.3 moles of gas C. There is no other gas present. The total pressure of the gases is 660 torr. What is true about the partial pressures of the gases?

(A) The partial pressure of gas A is 264 torr
(B) The partial pressure of gas B is 396 torr
(C) The partial pressure of gas C is 220 torr
(D) The partial pressures of gases A and C are each 198 torr
(E) The partial pressure of gas B is 330 torr

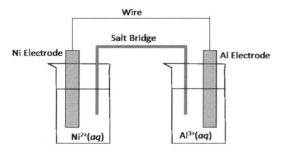

63. In the diagram above, what is the half reaction that occurs at the cathode?

(A) $Al \rightarrow Al^{3+} + 3e^-$
(B) $Ni^{2+} + 2e^- \rightarrow Ni$
(C) $Ni \rightarrow Ni^{2+} + 2e^-$
(D) $2Al^{3+} + 6e^- \rightarrow 2Al$
(E) $Al^{3+} + 3e^- \rightarrow Al$

64. In which of the following process, entropy likely increase?

(A) $Na(s) + H_2O\ (l) \rightarrow NaOH(aq) + H_2(g)$
(B) $I_2(g) \rightarrow I_2(s)$
(C) $H_2SO_4(aq) + Ba(OH)_2(aq) \rightarrow BaSO_4(s) + H_2O$
(D) $H_2O(g) \rightarrow H_2O(l)$
(E) None of above

65. The total amount of oxygen in an Al_2O_3 (molecular mass 102) sample is 5 moles. What is the total mass of the sample?

(A) 170 grams
(B) 255 grams
(C) 510 grams
(D) 765 grams
(E) 1020 grams

66. An alkaline earth metal, element M, reacts with oxygen. What is going to be the general formula for the compound formed?

(A) M_2O
(B) MO
(C) MO_2
(D) M_2O_3
(E) M_3O_2

67. Which functional group below does not contain a carbonyl group?

(A) Aldehydes
(B) Ketones
(C) Esters
(D) Ethers
(E) Carboxylic acids

68. $2Al(s) + 6HCl(aq) \rightarrow 2Al^{3+}(aq) + 6Cl^-(aq) + 3H_2(g)$

The reaction above is best described as a(n)

(A) Precipitation reaction.
(B) Acid/base reaction.
(C) Redox reaction.
(D) Combustion reaction.
(E) Decomposition reaction

GO ON TO THE NEXT PAGE

69. Which of the following statements is correct regarding a system at equilibrium?

   (A) The concentrations of reactants and products are the same.
   (B) The rates for forward and reverse reactions are the same.
   (C) Can be shifted by adding a catalysts.
   (D) Cannot be shifted by changing temperature or pressure.
   (E) Must always favor the formation of products.

70. What would be the stoichiometric coefficient of hydrochloric acid in the following equation?

   ....Cl_2 + ...H_2O → ...HCl + ...$HClO_3$

   (A) 1
   (B) 3
   (C) 5
   (D) 6
   (E) 10

STOP!

If you finish before time is called, you may check your work on this section only. Do not turn to any other section in the test.

Practice Test 5 Answers

PART A and C						PART B	
#	Answer	#	Answer	#	Answer	#	Answer
1	B	25	B	49	E	101	False, True, No
2	E	26	D	50	D	102	False, False, No
3	A	27	E	51	C	103	False, False, No
4	C	28	B	52	B	104	True, True, Yes
5	A	29	C	53	C	105	True, True, No
6	C	30	D	54	B	106	False, True, No
7	E	31	E	55	B	107	True, False, No
8	B	32	E	56	E	108	True, False, No
9	D	33	C	57	E	109	True, True, Yes
10	A	34	D	58	E	110	True, True, No
11	B	35	A	59	D	111	True, True, No
12	A	36	E	60	E	112	True, True, Yes
13	D	37	A	61	B	113	True, True, Yes
14	E	38	C	62	D	114	False, True, No
15	A	39	D	63	B	115	True, True, Yes
16	C	40	C	64	A	116	
17	B	41	A	65	A		
18	D	42	D	66	B		
19	E	43	E	67	D		
20	A	44	D	68	C		
21	C	45	A	69	B		
22	D	46	C	70	C		
23	A	47	B	71			
24	C	48	D	72			

Calculation of the raw score

The number of correct answers: _____ = No. of correct

The number of wrong answers: _____ = No. of wrong

Raw score = No. of correct − No. of wrong x ¼ = _____

Score Conversion Table

Raw Score	Scaled Score	Raw Score	Scaled Score	Raw Score	Scaled Score
80	800	49	600	18	420
79	800	48	590	17	410
78	790	47	590	16	410
77	780	46	580	15	400
76	770	45	580	14	390
75	770	44	570	13	390
74	760	43	560	12	380
73	760	42	560	11	370
72	750	41	550	10	360
71	740	40	550	9	360
70	740	39	540	8	350
69	730	38	540	7	350
68	730	37	530	6	340
67	720	36	520	5	340
66	710	35	520	4	330
65	700	34	510	3	330
64	700	33	500	2	320
63	690	32	500	1	320
62	680	31	490	0	310
61	680	30	490	−1	310
60	670	29	480	−2	300
59	660	28	480	−3	300
58	660	27	470	−4	290
57	650	26	470	−5	280
56	640	25	460	−6	280
55	640	24	450	−7	270
54	630	23	450	−8	270
53	620	22	440	−9	260
52	620	21	440	−10	260
51	610	20	430		
50	600	19	420		

Explanations: Practice Test 5

1. **(B)** Both boron and fluorine are nonmetal, and the B–F bond is polar covalent bond. However, the geometry of BF_3 molecule is trigonal planar since the central boron atom has sp^2 hybridization. Therefore, the BF_3 molecule is nonpolar.

2. **(E)** Strontium (Sr) is an alkali earth metal, and has metallic bonding.

3. **(A)** Potassium iodide (KI) is a typical ionic substance in which K atom transfers an electron to I atom to form oppositely charged ions, K^+ and I^-. The attraction between K^+ and I^- is electrostatic force.

4. **(C)** Formaldehyde, HCHO is a polar covalent substance as shown by the structure diagram $H-\overset{\overset{O}{\parallel}}{C}-H$.

5. **(A)** This formula is general structure of organic acids, which can be neutralized with a base.

6. **(C)** 2–pentanone ($\overset{\overset{O}{\parallel}}{}$) is a member of ketone which has general formula R–CO–R'.

7. **(E)** In chemistry, an alcohol is any organic compound in which the **hydroxyl** functional group (–OH) is bound to a carbon, with a general structural formula R-OH.

8. **(B)** An **aldehyde** (alkanal) is an organic compound containing a **functional group** –CHO, and general structure diagram $R-\overset{\overset{O}{\parallel}}{C}-H$.

9. **(D)** The molar mass of CO_2 is 44, and **88** grams of CO_2 is (88 gram)/(44 gram/mol) = **2.0 mol**.

10. **(A)** 1 mol CH_4 equals to **6.02 x 10^{23}** molecules by definition of Avogadro number.

11. **(B)** The molar mass of SO_2 is 64, and 32 grams of SO_2 equals to (32 gram)/(64 gram/mol) = 0.5 mol SO_2. At STP, 0.5 mole of ideal gas has a volume of (22.4 liter/mol) x 0.5 mol = **11.2 liter**.

12. **(A)** All alkali metals react with water vigorously to produce a strong base and hydrogen gas.

13. **(D)** Halogens have the greatest electronegativity in the same row (period). See **Figure** 2.2.

14. **(E)** Noble gases have stable outermost electron configuration, and highest **first ionization energy** in the same row (period).

15. **(A)** All alkali metals form X_2O with oxygen since they all have 1 valence electron.

16. **(C)** See **Table** 5.1.

Table 5.1 *pH Indicators*			
pH indicator	**Acid**	**pH range**	**Base**
Phenolphthalein	colorless	7.3 – 8.7	pink
Litmus	red	6.0 – 7.5	blue
Methyl red	red	4.8 – 6.2	yellow
Bromothymel	yellow	6.0 – 7.6	blue

17. **(B)** See **Table** 5.1.

18. **(D)** See **Table** 5.1.

19. **(E)** This is the expression of Dalton's law of partial pressure which states that total pressure is sum of partial pressure of all gases (**Table** 5.2).

20. **(A)** Boyle's law describes relationship between P and V when T is constant (**Table** 5.2).

Table 5.2 *Ideal Gas Laws*

Boyle's Law	Amonton's Law	Charle's Law	Avogadro's Law
$P_1V_1 = P_2V_2$	$P_1/T_1 = P_2/T_2$	$V_1/T_1 = V_2/T_2$	$V_1/n_1 = V_2/n_2$

Boyle's Law: If T = constant, PV = constant (V vs P)

Amonton's Law: If V = constant, T/P = constant (P vs T)

Charle's Law: If P = constant, V/T = constant (V vs T)

Avogadro's Law: If P, T = constant, V/n = constant (V vs n)

21. **(C)** This is the expression of general ideal gas law.

22. **(D)** This is the combined gas law (combining Boyle's, Amonton's and Charle's law) which relates gas parameters (P, V, and T).

23. **(A)** This is an alpha particle which is emitted when U–238 decays to Th–234.

24. **(C)** Gamma ray a high energy radiation (photon).

25. **(B)** This is a beta particle (electron) emitted when C–14 decay to N–14.

26. **(D)** Hydrocarbon has only carbon and hydrogen in the molecules, complete combustion of hydrocarbon requires that the products cannot be oxidized further by oxygen. H_2O and CO_2 are such products. Other choices have at least one compound which can be further oxidized.

27. **(E)** The completed reaction equation is $MgCO_3(s) + HCl(aq) \rightarrow MgCl_2(aq) + H_2O(l) + CO_2(g)$.

28. **(B)** Deviation of properties of real gas from ideal gas is caused by 2 assumptions of ideal gases: (1) The particles of ideal gas have no volume. For the real gases, the molecules do has a volume, and when the pressure keep increasing, the volume decreases to a point that the gas cannot be compressed further; (2) There is no interaction between particles of the gas except elastic collision. Intermolecular forces do exist in gases, and become increasingly important at low temperature and high pressure. When the volume of a gas decreases, the distance between particle decreases, and the interaction between particles increases.

At normal pressure and temperature, the intermolecular interaction is the main cause of the deviation of real gas from the ideal gas. The impact of molecule volume becomes significant only when the real gas compressed to very small volume at low temperature. Therefore, choice B is the best answer.

29. **(C)** This problem is about balancing redox reaction equation under acidic condition. For such reaction, not only all atoms need to be balanced, but also the transfer of electrons is balanced. In addition, H^+ or water are needed to balance H and O. **Table** 5.3 list steps to balance such equations. Also see **Table** 4.1 for balancing redox equation without acid or base involved.

Table 5.3 *Balance Redox Reaction Under Acidic Condition*

Balance redox reaction equation: $Ce^{4+} + Bi \rightarrow Ce^{3+} + BiO^+$.

Step 1: Separate the half–reactions.

$Ce^{4+}(aq) + e^- \rightarrow Ce^{3+}(aq)$ (1)

$$Bi(s) \rightarrow BiO^+ (aq) + 3e^- \quad (2)$$

Step 2: Balance the half reactions. Add water and H^+/OH^- if necessary.

$$Ce^{4+}(aq) + e^- \rightarrow Ce^{3+} (aq) \quad (1)$$
$$Bi(s) + H_2O \rightarrow BiO^+ (aq) + 2H^+ + 3e^- \quad (3)$$

Step 3: Balance the electrons in the half equations. In this case, half equation (1) is multiplied by 3 and half equation (2) is not changed. Numbers of gain or loss of electrons are balanced.

$$3Ce^{4+}(aq) + 3e^- \rightarrow 3Ce^{3+} (aq) \quad (4)$$
$$Bi(s) + H_2O \rightarrow BiO^+ (aq) + 2H^+ + 3e^- \quad (5)$$

Step 3: Adding the half equations:

$$3Ce^{4+}(aq) + 3e^- + Bi(s) + H_2O \rightarrow 3Ce^{3+} (aq) + BiO^+ (aq) + 2H^+ + 3e^-$$

Step 4: The electrons cancel out:

$$3Ce^{4+}(aq) + Bi(s) + H_2O \rightarrow 3Ce^{3+} (aq) + BiO^+ (aq) + 2H^+$$

30. **(D)** Molar mass of oxygen gas, O_2, is 32 grams/mol. At STP, 1 mol of O_2 occupies 22.4 liter of space. The density of O_2 at STP (32 gram)/(22.4 liter) = 1.43 gram/L.

Alternatively, you can solve this problem using the general ideal gas law as above. If you remember the formula for calculating gas density from molar mass, you can do the following:

$$d = PM/RT = (1 \text{ atm x } 32 \text{ g mol}^{-1})/(0.0821 \text{ atm L mol}^{-1} K^{-1} \text{ x } 273 \text{ K}) = 1.43 \text{ g/L}$$

31. **(E)** is correct answer.

(A) Water has a bent molecular geometry and ~~one~~ (**two**) lone pair of electrons.
(B) Ammonia has a trigonal pyramidal molecular geometry and ~~two~~ (**one**) lone pairs of electrons.
(C) Methane has a ~~trigonal planar~~ (**tetrahedral**) molecular geometry.
(D) Carbon dioxide is linear because it has ~~one single bond and one triple bond~~ (**two double bonds**).

32. **(E)** Assume the molar mass of the compound is M, then the molar ratio
C : Cl : O = (M x 12.1%/12) : (M x 71.7%/36.5) : (M x 16.2%//16) = 1 : 2 : 1.
The empirical of the compound is CCl_2O.

33. **(C)** Phosphorus has atomic number of 15. The numbers of electrons in first to third layer are 2, 8, 5 respectively.

34. **(D)** Adding catalyst and increasing temperature both increase reaction rate (**Table** 1.7). However, adding He has no impact on reaction rate.

35. **(A)** Halogens has 7 valence electrons, and tend to gain one electron to achieve octet state and form X^- ion. All other statements are incorrect as

(B) They have the ~~lowest~~ (**highest**) ionization energy of the elements in the respective period.
(C) They are ~~all gases~~ at room temperature. (Br_2 is liquid, and I_2 is solid at room temperature)
(D) They are among the best ~~reducing~~ (**oxidizing**) reagents.
(E) They all react with water to form ~~basic~~ (**acidic**) solution.

36. **(E)** Xenon is the biggest in the listed noble gases, and the interaction between xenon atoms are also strongest. Therefore, Xe shows largest deviation from ideal gas behavior.

37. **(A)** Refer to **Table** 1.4 for freezing point depression and **Table** 2.6 for boiling point elevation.

$Cu(NO_3)_2$ and NaBr solutions have the same molality (1.0 m), but Van't Hoff factor i of $Cu(NO_3)_2$ and NaBr are 3 vs. 2 respectively. Therefore, 1.0 m $Cu(NO_3)_2$ will cause greater boiling point elevation and freezing point depression.

38. **(C)** Referring to the heating diagrams below, heat needed to complete the process is divided to two parts: heat to melt the ice (B to C) and heat to change temperature of liquid water from 0°C to 100°C (C to D).

Heat to melt ice: $\Delta H_{fusion} = 10$ g x 80 cal/g = 800 cal

Heat to change temperature of liquid water from 0°C to 100°C:

$\Delta H_{water} = 10$ g x 1.00 cal/g °C x (100°C – 0°C) = 1000 cal.

Total heat: $\Delta H = 800$ cal + 1000 cal = 1800 cal.

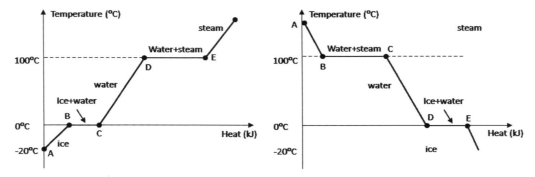

Figure 5.1 Phase change diagram of water in heating (left) and cooling (right) processes.

39. **(D)** HCl is gas at normal temperature and pressure, and its solubility decrease with increase in temperature.

40. **(C)** At STP, 67.2 liter CO_2 = (67.2 L CO_2)/(22.4 L CO_2/mol CO_2) = 3.0 mol CO_2.

Since the ratio of consumption of CH_4 to production of CO_2 is 1:1, there are 3 mol of CH_4 is consumed. Mass of CH_4 burn: 3 mol CH_4 x 18 gram CH_4/mol CH_4= 54 grams CH_4.

41. **(A) La Chatelier's principle** states that if a chemical system at equilibrium experiences a change in concentration, temperature, volume, or partial pressure, then the equilibrium shifts to counteract the imposed change and a new equilibrium is established. When the temperature increases, the system will shift to the direction at which heat is absorbed to cancel the increase in temperature, i.e. the reaction shift to forward direction (**Table** 1.5). This leads to more product and less reactants.

42. **(D)** $K_{eq} = \frac{[NO_2Cl]^2}{[Cl_2][NO_2]^2} = \frac{0.5^2}{0.3 \times 0.2^2} = 20.8$

Note: although the concentrations in the equilibrium have units, the K_{eq} is dimensionless.

43. **(E)** N in NH_3 has sp^3 hybridization, and electron configuration is tetrahedral. Since N has 5 valence electrons, one sp^3 hybridized orbital is filled with a pair of electrons, and the other three sp^3 orbital have one electron each and form covalent bond with one H. The shape of NH_3 molecule is **trigonal pyramid**.

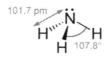

44. **(D)** A solution with [OH⁻] = 10^{-3} M is a base. Only D is a strong base and after 0.001 M NaOH is completely dissociated, [OH⁻] = 10^{-3} M. Choices A and C are weak acids; B is strong acid; E is neutral salt.

45. **(A)** At STP, 1 mol of the gas (6.02 x 10^{23} molecules) has a volume of 22.4 liter.

The molar mass of the gas is 64.0 gram/mol (this gas likely is SO_2).
The number of mol of 8 grams of this gas is 8/64 = 0.125 mol.
The volume of 8 grams of this gas at STP is: (22.4 liter/mol) x (0.125 mol) = 2.8 liter.

46. **(C)** is the correct answer. $CH_3CH(Cl)CH_3$ and $CH_3CH_2CH_2Cl$ have the same formula, i.e. C_3H_7Cl, but they are different compounds. The difference is the position of Cl. In $CH_3CH(Cl)CH_3$, Cl is connected to the middle C atom, while in $CH_3CH_2CH_2Cl$, Cl is connected to one end C.

    (A) $CH_3CH_2CH_2OH$ and $HOCH_2CH_2CH_3$ are virtually the same compounds.
    (B) $CH_3CH_2CH_3$ and $CH_3CH_2CH_2CH_3$ have different formula (C_3H_8 and C_4H_{10})
    (C) $CH_3CH(Cl)CH_3$ and $CH_3CH_2CH_2Cl$ are isomers
    (D) CH_3COCH_3 and $CH_3CH_2CH_2CHO$ have different formula (C_3H_6O and C_4H_9O)
    (E) $ClCH_2CH_2Br$ and $BrCH_2CH_2Cl$ are the same compounds.

47. **(B)** Isotopes of an element have the same atomic (proton) number but different mass (neutron) numbers. Choices B meets this definition.

	Proton	Electron	Neutron	*Explanation*
(A)	18	19	19	This is the ion (−1) of the same atom.
(B)	**18**	**18**	**18**	**Isotope of the atom with one less neutron**
(C)	17	17	17	Different element
(D)	19	18	19	Different element
(E)	19	19	18	Different element

48. **(D)** The balanced equation is _2_ $C_2H_6(g)$ + _7_ $O_2(g)$ → _4_ $CO_2(g)$ + _6_ $H_2O(g)$.

49. **(E)** Supersaturated solution contains more solute than solvent can dissolve under normal circumstances. One example is when a saturated solution is cooled down, the solubility decreases with temperature, but the solute remains in the solution. In this case, the solution become supersaturated not because more solute is dissolved in solvent after it reaches saturation (this is usually impossible), but because the decrease in solubility. A supersaturated solution is very unstable, a small change can trigger the crystallization process. One such trigger is adding a small crystal of the salt in the solution, which will serve as a center for crystallization, this process will proceed until the solution reach saturation again. At saturation the solid phase of solute and the solution reaches equilibrium.

50. **(D)** Molar mass of $Ca(OH)_2$, KCl and LiBr are 57.0, 74.5 and 87.0 respectively. The concentration of solution I, II and III are 1.3 M, 1.0 M and 1.0 M respectively.

51. **(C)** Calculate mass of H_2 and N_2 as follow:

    Step 1: Number of mol of 68 gram of NH_3: (68 gram)/(17 gram/mol) = 4 mol NH_3.

    Step 2: Number of mol of H_2: 4 mol NH_3 x (3 mol H_2/2 mol NH_3) = 6 mol H_2.
    Number of mol of N_2: 4 mol NH_3 x (1 mol N_2/2 mol NH_3) = 2 mol N_2.

    Step 3: Mass of H_2: (6 mol H_2) x (2 gram/mol) = 12 gram H_2.
    Mass of N_2: (2 mol N_2) x (28 gram/mol) = 56 gram N_2.

52. **(B)** C_2H_4 is ethene which has one carbon-carbon double bond, C=C. C_2H_2 is ethyne which contains a triple bond, C≡C. C_3H_8 and C_4H_{10} are all saturated hydrocarbon. C_2H_6O is an ether, CH_3–O–CH_3, and no double bond in it.

53. **(C)** Neutralization reaction between HNO_3 (a strong acid) and KOH (a strong base) produces water and a salt (KNO_3) as in the equation: HNO_3 + KOH → **KNO_3** + H_2O

54. **(B)** After adding 200 mL water, total volume of the solution increases to 600 mL. Use the law of mass conservation: $M_1V_1 = M_2V_2$. Rearrange to solve the concentration after dilution M_2:

$M_2 = M_1V_1/V_2 = (0.6 \text{ M})(400 \text{ mL})/(600 \text{ mL}) = 0.4 \text{ M}.$

55. **(B)** is correct. Salt and water can be separated by distillation since water is evaporated while salt will remain as solid substance.

   (A) Oil and water— ~~filter paper~~ separatory funnel. Oil and water separate to layers which can be separate with separatory funnel.
   (B) Correct answer.
   (C) Sand and water— ~~separatory funnel~~, filter paper. Sand is solid, can be retained on filter paper while water going through filter paper.
   (D) Salt and sugar— ~~filter paper~~. Both soluble in water, and cannot be separated with filter paper. There is no simplistic method to separate salt from sugar.
   (E) Sugar and water— ~~centrifuge~~ distillation. Similar to Choice B.

56. **(E)** is correct answer because RbI is soluble. Referring to the solubility rules in **Table** 2.3, AgCl, $Pb_3(PO_4)_2$, $CaCO_3$, Hg_2Br_2 are insoluble and reaction between the two ions form precipitation. Rb is a alkali metal and RbI is soluble.

57. **(E)** Referring to **Le Chatelier's Principle (Table** 1.5), when Na_2SO_4 is added, concentration of SO_4^{2-} increases, the equilibrium will shift to left (**E is correct answer**). A and D are incorrect since the solubility and K_{sp} do not change when salt is added (temperature can change the solubility and K_{sp}). B is incorrect since when the equilibrium shift to left, Ca^{2+} concentration will decrease.

58. **(E)** Formula pf iron (III) chloride is $FeCl_3$. Hexahydrate indicates that there are 6 hydrate, and write as $6H_2O$ in the overall formula. And finally, use interpunct "•" between the salt and hydrate water as: $FeCl_3 \cdot 6H_2O$.

59. **(D)** Only methane does not react with NaOH.

60. **(E)** Iodine, carbon dioxide and naphthalene are three best examples which sublime at normal temperature.

61. **(B)** The balanced reaction equation is $Pb(NO_3)_2 + 2NaI \rightarrow PbI_2 + 2NaNO_3$. The molar ratio of $Pb(NO_3)_2$ to NaI and to PbI_2 is 1:2:1. The first step is to determine which reactant is limiting, second step is to determine which mixture yields more PbI_2 precipitate.

Mixture	mL of 0.2 M $Pb(NO_3)_2$	mL of 0.2 M NaI	Limiting reactant	PbI₂ produced
(A)	2	7	$Pb(NO_3)_2$	2 mL x 0.2 M
(B)	3	6	None	3 mL x 0.2 M
(C)	4	5	NaI	2.5 mL x 0.2 M
(D)	5	4	NaI	2 mL x 0.2 M
(E)	6	3	NaI	1.5 mL x 0.2 M

62. **(D)** The total number of mol of gases: 0.3 mol + 0.4 mol + 0.3 mol = 1.0 mol.

   Partial pressure of A and C is (660 torr) x 0.3/1 = 198 torr, and the partial pressure of B is (660 torr) x 0.4 mol/1.0 mol = 264 torr.

63. **(B)** Based on the reactivity series, Al is more reactive than Ni. Al is oxidized (anode), and Ni^{2+} is reduced (cathode) as expressed in the equation $Al + Ni^{2-} \rightarrow Al^{3+} + Ni$. In a voltaic cell, the oxidation reaction occurs on anode, and reduction reaction occurs on **cathode ($Ni^{2+} + 2e^- \rightarrow Ni$)**.

64. **(A)** is correct answer. From pure solid sodium and liquid water to sodium hydroxide solution and hydrogen gas, the system become less orderly (more random), i.e. entropy increases.

For both B and D, when gas become solid or liquid, the randomness (entropy) of system decrease. For C, when precipitation (solid) is formed, the system also become less random, entropy decreases too.

65. **(A)** There are 3 mol of O in 1 mol Al_2O_3;
Number of mol of Al_2O_3: 5 mol O x (1 mol Al_2O_3 / 3 mol O) = 1.67 mol Al_2O_3;
Mass of sample: 1.67 mol Al_2O_3 x 102 g/mol = 170 g Al_2O_3.

66. **(B)** Alkaline earth metals (Be, Mg, Ca etc.) belong to IIA group and all have 2 valence electron, and form cation with +2 charge, oxygen usually gain two electrons to form anion with –2 charge. Therefore, formula has 1:1 ration for alkaline earth metals and oxygen.

67. **(D)** A carbonyl group is a functional group composed of a carbon atom double–bonded to an oxygen atom: C=O. Aldehyde, ketone, carboxylic acid, ester all have at least one carbonyl group, while ether has one oxygen atom to connect to two carbon atoms through single bond (C–O–C).

68. **(C)** This is a reduction–oxidation reaction, Al is oxidized and H^+ is reduced.

69. **(B)** At chemical equilibrium, the rates of forward and reverse reaction are the same (B is correct); and the concentration of reactants and products remain constant (but not the same, A is incorrect). Change in pressure or temperature may shift equilibrium depending on the situation (D is incorrect). Catalyst increase reaction rates for both forward and reverse reactions, but does not shift equilibrium (C is incorrect). At equilibrium, neither reactants nor products are favored (E is incorrect).

70. **(C)** The balanced equation is: _3_ Cl_2 + _3_ H_2O → _5_ HCl + _1_ $HClO_3$.

This is reduction–oxidation reaction, and electron transfer need to be balanced. This reaction is unique in that Cl_2 play roles of both reducing and oxidizing reagents.

101. **Correct answer: I False, II True, CE No**

Explanation: ^{14}C and ^{14}N belong to different elements, they are not isotopes to each other (Statement I is false). They don't have same number of protons (atomic number), but they do have same mass number (Statement II is true).

102. **Correct answer: I True, II True, CE No**

Explanation: Ice is less dense than liquid water due to the crystalline structure in ice. Hydrogen bonds can be found in both liquid and solid water; but in ice, the molecules of water are aligned in more ordered manner, and the water molecules are held together by hydrogen bonds more tightly. In liquid water, hydrogen bonds are loose and can be broken easily. This difference causes the difference in the density. The polarity of water molecules does not explain the difference in density, statement II is true but does not explain statement I.

103. **Correct answer: I False, II False, CE No**

Explanation: A solution with a pH of 5 has concentration of H^+ ion 10^{-5} M, which is 1000 times higher than the concentration of H^+ ion in a solution with pH of 8 ($[H^+] = 10^{-8}$ M). Solution of pH of 5 is more acidic than that of pH of 8 (both statements I and II are false).

104. **Correct answer: I True, II True, CE Yes**

Explanation: Transition metals have partially filled d orbitals (Statement II is true), this explains why valence electrons in transition metals can be excited easily by absorbing visible light, and display color (Statement I is true). And Statement II correctly explains Statement I.

105. **Correct answer: I True, II True, CE No**

Explanation: In a voltaic cell, the reduction–oxidation reaction proceeds spontaneously because the two metals involved in the reactions have different redox potential (Statement I is true). For example, Zn can react with $CuSO_4$ spontaneously in the reaction $Zn(s) + CuSO_4(aq) \rightarrow ZnSO_4(aq) + Cu(s)$. A voltaic cell with Zn (anode) and Cu (cathode) as electrodes will generate electric current spontaneously. The reason to put Zn and Cu in different containers is to harness the electric current (Statement II is true), but this does not explain why the reaction is spontaneously (CE is No).

106. **Correct answer: I False, II True, CE No**

Explanation: Although F^- and Ne has the same number of electrons and electron configuration (Statement II is true), their chemical properties are very different (Statement I is false). F^- is negatively charged and can be involved in many chemical reaction, while Ne is a noble gas and chemically inactive.

107. **Correct answer: I True, II False, CE No**

Explanation: At equilibrium, the reaction rates of forward and reverse reactions are the same, but not zero (Statement II is false); that's why the concentration of reactant and products remain unchanged (Statement I is true).

108. **Correct answer: I True, II False, CE No**

Explanation: Elements in the same group have similar chemical properties (Statement I is true) because they all have the same number of valence electrons, and similar electronic configuration in the valence shell. But this does not mean that their valence electrons has the same energy (Statement II is false).

109. **Correct answer: I True, II True, CE Yes**

Explanation: n–butane and 2–methylpropane are saturated hydrocarbon with same formula (C_4H_{10}, Statement II is true), they are different in their structure, n–butane is straight–chained and 2–methylpropane has a branch. They fit the definition of isomer (Statement I is true), and Statement II correctly explains Statement I.

110. **Correct answer: I True, II True, CE No**

Explanation: Fluorine and chlorine are in the same group, chlorine has one more electron shell and is larger than fluorine (statement II is true). But atom size solely does not explain why fluorine is stronger oxidizing reagent, greater affinity to electron by F than Cl explains why F is stronger oxidizing reagent.

111. **Correct answer: I True, II True, CE No**

Explanation: Neutral chlorine (Cl) and potassium (K) atoms have 17 and 19 electrons respectively. Cl tends to gain one electron to become Cl^- ion with one negative charge, and K tends to lose one electron to become K^+ ion. Both Cl^- and K^+ ions have 18 electrons, which is the same as argon atom (Statement II is true). The Cl^- and K^+ ions can form ionic bond, because the difference of electronegativity between chlorine and potassium is large (Statement I is true). However, the fact that Cl^- and K^+ has same electron configuration has nothing with the ionic bond between Cl^- and K^+.

112. **Correct answer: I True, II True, CE Yes**

Explanation: When ice melts, the crystalline structure in ice is broken, water molecules in liquid can move more freely than in ice. The system become less organized, i.e. entropy of the system increases. Both statements are true, and Statement II correctly explains Statement I.

113. **Correct answer: I True, II True, CE Yes**

Explanation: Nitrogen atom has 5 valence electrons, it forms three N–H bonds through sp^3 hybridization. One lone pair of electron repel the bonding electron of the 3 N–H bond, and form the trigonal pyramidal geometry rather than trigonal planar geometry. Both statements are true, and Statement II correctly explains Statement I.

114. **Correct answer: I False, II True, CE No**

Explanation: The proper name of $AlCl_3$ is aluminum chloride, since the Al–Cl bond is ionic bond and the prefix "tri–" is not used. Statement II is true anyway, but does not apply to $AlCl_3$.

115. **Correct answer: I True, II True, CE Yes**

Explanation: The combustion of fossil fuel releases carbon oxides along with some other minor oxide gases such as sulfur oxide and nitrogen oxide. These oxides gas can dissolved in water to form the acidic precipitation, and this is the main cause of acid rain. Both statements are true, and Statement II correctly explains Statement I.

SAT II Chemistry

Practice Test 6

Answer Sheet

Part A and C: Determine the correct answer. Blacken the oval of your choice completely with a No. 2 pencil.

1	Ⓐ Ⓑ Ⓒ Ⓓ Ⓔ	25	Ⓐ Ⓑ Ⓒ Ⓓ Ⓔ	49	Ⓐ Ⓑ Ⓒ Ⓓ Ⓔ
2	Ⓐ Ⓑ Ⓒ Ⓓ Ⓔ	26	Ⓐ Ⓑ Ⓒ Ⓓ Ⓔ	50	Ⓐ Ⓑ Ⓒ Ⓓ Ⓔ
3	Ⓐ Ⓑ Ⓒ Ⓓ Ⓔ	27	Ⓐ Ⓑ Ⓒ Ⓓ Ⓔ	51	Ⓐ Ⓑ Ⓒ Ⓓ Ⓔ
4	Ⓐ Ⓑ Ⓒ Ⓓ Ⓔ	28	Ⓐ Ⓑ Ⓒ Ⓓ Ⓔ	52	Ⓐ Ⓑ Ⓒ Ⓓ Ⓔ
5	Ⓐ Ⓑ Ⓒ Ⓓ Ⓔ	29	Ⓐ Ⓑ Ⓒ Ⓓ Ⓔ	53	Ⓐ Ⓑ Ⓒ Ⓓ Ⓔ
6	Ⓐ Ⓑ Ⓒ Ⓓ Ⓔ	30	Ⓐ Ⓑ Ⓒ Ⓓ Ⓔ	54	Ⓐ Ⓑ Ⓒ Ⓓ Ⓔ
7	Ⓐ Ⓑ Ⓒ Ⓓ Ⓔ	31	Ⓐ Ⓑ Ⓒ Ⓓ Ⓔ	55	Ⓐ Ⓑ Ⓒ Ⓓ Ⓔ
8	Ⓐ Ⓑ Ⓒ Ⓓ Ⓔ	32	Ⓐ Ⓑ Ⓒ Ⓓ Ⓔ	56	Ⓐ Ⓑ Ⓒ Ⓓ Ⓔ
9	Ⓐ Ⓑ Ⓒ Ⓓ Ⓔ	33	Ⓐ Ⓑ Ⓒ Ⓓ Ⓔ	57	Ⓐ Ⓑ Ⓒ Ⓓ Ⓔ
10	Ⓐ Ⓑ Ⓒ Ⓓ Ⓔ	34	Ⓐ Ⓑ Ⓒ Ⓓ Ⓔ	58	Ⓐ Ⓑ Ⓒ Ⓓ Ⓔ
11	Ⓐ Ⓑ Ⓒ Ⓓ Ⓔ	35	Ⓐ Ⓑ Ⓒ Ⓓ Ⓔ	59	Ⓐ Ⓑ Ⓒ Ⓓ Ⓔ
12	Ⓐ Ⓑ Ⓒ Ⓓ Ⓔ	36	Ⓐ Ⓑ Ⓒ Ⓓ Ⓔ	60	Ⓐ Ⓑ Ⓒ Ⓓ Ⓔ
13	Ⓐ Ⓑ Ⓒ Ⓓ Ⓔ	37	Ⓐ Ⓑ Ⓒ Ⓓ Ⓔ	61	Ⓐ Ⓑ Ⓒ Ⓓ Ⓔ
14	Ⓐ Ⓑ Ⓒ Ⓓ Ⓔ	38	Ⓐ Ⓑ Ⓒ Ⓓ Ⓔ	62	Ⓐ Ⓑ Ⓒ Ⓓ Ⓔ
15	Ⓐ Ⓑ Ⓒ Ⓓ Ⓔ	39	Ⓐ Ⓑ Ⓒ Ⓓ Ⓔ	63	Ⓐ Ⓑ Ⓒ Ⓓ Ⓔ
16	Ⓐ Ⓑ Ⓒ Ⓓ Ⓔ	40	Ⓐ Ⓑ Ⓒ Ⓓ Ⓔ	64	Ⓐ Ⓑ Ⓒ Ⓓ Ⓔ
17	Ⓐ Ⓑ Ⓒ Ⓓ Ⓔ	41	Ⓐ Ⓑ Ⓒ Ⓓ Ⓔ	65	Ⓐ Ⓑ Ⓒ Ⓓ Ⓔ
18	Ⓐ Ⓑ Ⓒ Ⓓ Ⓔ	42	Ⓐ Ⓑ Ⓒ Ⓓ Ⓔ	66	Ⓐ Ⓑ Ⓒ Ⓓ Ⓔ
19	Ⓐ Ⓑ Ⓒ Ⓓ Ⓔ	43	Ⓐ Ⓑ Ⓒ Ⓓ Ⓔ	67	Ⓐ Ⓑ Ⓒ Ⓓ Ⓔ
20	Ⓐ Ⓑ Ⓒ Ⓓ Ⓔ	44	Ⓐ Ⓑ Ⓒ Ⓓ Ⓔ	68	Ⓐ Ⓑ Ⓒ Ⓓ Ⓔ
21	Ⓐ Ⓑ Ⓒ Ⓓ Ⓔ	45	Ⓐ Ⓑ Ⓒ Ⓓ Ⓔ	69	Ⓐ Ⓑ Ⓒ Ⓓ Ⓔ
22	Ⓐ Ⓑ Ⓒ Ⓓ Ⓔ	46	Ⓐ Ⓑ Ⓒ Ⓓ Ⓔ	70	Ⓐ Ⓑ Ⓒ Ⓓ Ⓔ
23	Ⓐ Ⓑ Ⓒ Ⓓ Ⓔ	47	Ⓐ Ⓑ Ⓒ Ⓓ Ⓔ	71	Ⓐ Ⓑ Ⓒ Ⓓ Ⓔ
24	Ⓐ Ⓑ Ⓒ Ⓓ Ⓔ	48	Ⓐ Ⓑ Ⓒ Ⓓ Ⓔ	72	Ⓐ Ⓑ Ⓒ Ⓓ Ⓔ

Part B: On the actual Chemistry Test, the following type of question must be answered on a special section (labeled "Chemistry") at the lower left–hand corner of your answer sheet. These questions will be numbered beginning with 101 and must be answered according to the directions.

PART B	I		II		CE
101	T	F	T	F	◯
102	T	F	T	F	◯
103	T	F	T	F	◯
104	T	F	T	F	◯
105	T	F	T	F	◯
106	T	F	T	F	◯
107	T	F	T	F	◯
108	T	F	T	F	◯
109	T	F	T	F	◯
110	T	F	T	F	◯
111	T	F	T	F	◯
112	T	F	T	F	◯
113	T	F	T	F	◯
114	T	F	T	F	◯
115	T	F	T	F	◯
116	T	F	T	F	◯

Periodic Table of Elements

Material in this table may be useful in answering the questions in this examination

1 H 1.0079																	2 He 4.0026
3 Li 6.941	4 Be 9.012											5 B 10.811	6 C 12.011	7 N 14.007	8 O 16.00	9 F 19.00	10 Ne 20.179
11 Na 22.99	12 Mg 24.30											13 Al 26.98	14 Si 28.09	15 P 30.974	16 S 32.06	17 Cl 35.453	18 Ar 39.948
19 K 39.01	20 Ca 40.48	21 Sc 44.96	22 Ti 47.90	23 V 50.94	24 Cr 52.00	25 Mn 54.938	26 Fe 55.85	27 Co 58.93	28 Ni 58.69	29 Cu 63.55	30 Zn 65.39	31 Ga 69.72	32 Ge 72.59	33 As 74.92	34 Se 78.96	35 Br 79.90	36 Kr 83.80
37 Rb 85.47	38 Sr 87.62	39 Y 88.91	40 Zr 91.22	41 Nb 92.91	42 Mo 95.94	43 Tc (98)	44 Ru 101.1	45 Rh 102.91	46 Pd 106.42	47 Ag 107.87	48 Cd 112.41	49 In 114.82	50 Sn 118.71	51 Sb 121.75	52 Te 127.60	53 I 126.91	54 Xe 131.29
55 Cs 132.91	56 Ba 137.33	57 *La 138.91	72 Hf 178.49	73 Ta 180.95	74 W 183.85	75 Re 186.21	76 Os 190.2	77 Ir 192.2	78 Pt 195.08	79 Au 196.97	80 Hg 200.59	81 Tl 204.38	82 Pb 207.2	83 Bi 208.98	84 Po (209)	85 At (210)	86 Rn (222)
87 Fr (223)	88 Ra 226.02	89 Ac 227.03	104 Rf (261)	105 Db (262)	106 Sg (266)	107 Bh (264)	108 Hs (277)	109 Mt (268)	110 Ds (271)	111 Rg (272)	112 (277)						

	58 Ce 140.12	59 Pr 140.91	60 Nd 144.24	61 Pm (145)	62 Sm 150.4	63 Eu 151.97	64 Gd 157.25	65 Tb 158.93	66 Dy 162.50	67 Ho 164.93	68 Er 167.26	69 Tm 168.93	70 Yb 173.04	71 Lu 174.97
*Lanthanide Series														
Actinide Series	90 Th 232.04	91 Pa 231.04	92 U 238.03	93 Np 237.05	94 Pu (244)	95 Am (243)	96 Cm (247)	97 Bk (247)	98 Cf (251)	99 Es (252)	100 Fm (257)	101 Md (258)	102 No (259)	103 Lr (260)

Note: For all questions involving solutions, assume that the solvent is water unless otherwise stated.
Reminder: You may not use a calculator in this test!

Throughout the test the following symbols have the definitions specified unless otherwise noted.

H = enthalpy	atm = atmosphere(s)
M = molar	g = gram(s)
n = number of moles	J = joule(s)
P = pressure	kJ = kilojoule(s)
R = molar gas constant	L = liter(s)
S = entropy	mL = milliliter(s)
T = temperature	mm = millimeter(s)
V = volume	mol = mole(s)
	V = volt(s)

Chemistry Subject Practice Test 6

Part A

Directions for Classification Questions
Each set of lettered choices below refers to the numbered statements or questions immediately following it. Select the one lettered choice that best fits each statement or answers each question and then fill in the corresponding circle on the answer sheet. <u>A choice may be used once, more than once, or not at all in each set.</u>

Questions 1–5 refer to the following pairs

    (A) Al and S
    (B) Li^+ and Na^+
    (C) Be^{2+} and H
    (D) N and P^{3-}
    (E) Ca^{2+} and Cl^-

1. have the same number of electrons.

2. differ in their number of electrons by 1.

3. differ in their number of electrons by 3.

4. differ in their number of electrons by 8.

5. differ in their number of electrons by 11.

Questions 6–9 refer to the following compounds

    (A) NH_3
    (B) H_2O
    (C) BeF_2
    (D) BCl_3
    (E) CH_4

6. has linear geometry.

7. has tetrahedral geometry.

8. has trigonal pyramidal geometry.

9. has trigonal planar geometry.

Questions 10–13 refer to the following phase diagram. The horizontal arrows refer to phase changes as the temperature changes at constant pressure.

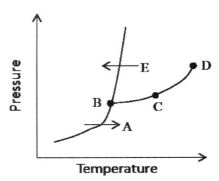

10. The temperature and pressure condition under which sold, liquid and gas can exist simultaneously

11. The point beyond which pressure alone cannot liquefy the gas.

12. The process known as sublimation.

13. The temperature and pressure condition under which liquid and vapor are at equilibrium.

GO ON TO THE NEXT PAGE

Questions 14–17 refer to the heating curve of a substance

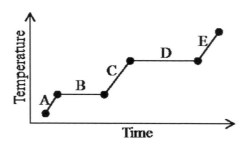

14. Represents vaporization of liquid.

15. Represent melting of solid.

16. The liquid state is heated from freezing point to boiling point.

17. The solid state is heated to melting point.

Questions 18–21 refer to the following types of solutions

    (A) Buffer solution
    (B) Strong acid
    (C) Weak acid
    (D) Strong base
    (E) Weak base

18. $Ba(OH)_2$ solution

19. H_3PO_4 solution

20. CH_3COOH solution

21. Ammonia solution

Questions 22–25 refer to the following laboratory procedures:

    (A) Distillation
    (B) Chromatography
    (C) Fractional crystallization
    (D) Filtration
    (E) Titration

22. is used to separate a precipitate from a solution.

23. is used to separate a mixture of liquids based on differences in their boiling points.

24. is used to determine the unknown concentration of acid/base using base/acid of known concentration

25. can be used to separate a mixture of dissolved solids based on difference in their solubility.

GO ON TO THE NEXT PAGE

PLEASE GO TO THE SPECIAL SECTION AT THE LOWER LEFT–HAND CORNER OF PAGE 2 OF YOUR ANSWER SHEET LABELED CHEMISTRY AND ANSWER QUESTIONS 101–115 ACCORDING TO THE FOLLOWING DIRECTIONS.

Part B

Directions for Relationship Analysis Questions
Each question below consists of two statements, I in the left–hand column and II in the right–hand column. For each question, determine whether statement I is true or false and whether statement II is true or false and fill in the corresponding T or F circles on your answer sheet. *Fill in circle CE only if statement II is a correct explanation of the true statement I.*

EXAMPLES:

I		II
EX1. The nucleus in an atom has a positive charge.	BECAUSE	Proton has positive charge, neutron has no charge.

SAMPLE ANSWERS

	I		II		CE
EX1	⬤	Ⓕ	⬤	Ⓕ	⬤

	I		II
101.	When 1.0 mol H_2SO_4 reacts with 1.0 mol NaOH, NaOH is the limiting reactant	BECAUSE	the molar mass of H_2SO_4 is more than twice the molar mass of NaOH.
102.	At low temperature and high pressure, gases tend to condense	BECAUSE	gases tend to fill the container in which they are placed.
103.	H_2O and H_2O_2 have the same empirical formula	BECAUSE	in H_2O and H_2O_2, oxygen has the same oxidation state.
104.	Bromine is a stronger oxidizing reagent than chlorine	BECAUSE	bromine atoms are larger than chlorine atoms.
105.	Combustion of hydrocarbon is a exothermic reaction	BECAUSE	heat must be added to the system for the exothermic reaction.
106.	A molecule of silicon tetrachloride, $SiCl_4$, is nonpolar	BECAUSE	the four bonds in $SiCl_4$ are identical and the molecular has a tetrahedral structure.

107. The temperature of a substance always increase as heat is added to it — BECAUSE — The average kinetic energy of the particles in a system increases as temperature increase.

108. When magnesium reacts with chlorine, the atoms combine in a 1 to 2 ratio to form $MgCl_2$. — BECAUSE — Each magnesium atom gains two electrons and each chlorine atom loses one electron.

109. Atomic radii decrease down a group — BECAUSE — the atomic number increase downward within a group.

110. When an ideal gas is cooled its volume will increase — BECAUSE — when pressure of a gas decreases at constant temperature, it expands.

111. Nitrogen gas effuses faster than oxygen gas — BECAUSE — the diatomic molecules of nitrogen are lighter than those of oxygen.

112. The reaction $Al^{3+} + 3e^- \rightarrow Al$ is reduction reaction. — BECAUSE — in reduction reaction, electron(s) is(are) gained.

113. Water makes a good buffer — BECAUSE — water molecules are amphoteric.

114. Carbon tetrachloride is insoluble in water — BECAUSE — water molecules contain polar bonds and have an asymmetric geometry.

115. The entropy of a solid increases when it is dissolved in solvent — heat is always released when a solid dissolved in solvent.

GO ON TO THE NEXT PAGE

Part C

Directions for Five–Choice Completion Questions
Each of the questions or incomplete statements below is followed by five suggested answers or completions. Select the one that is best in each case and then fill in the corresponding circle on the answer sheet.

26. 50 mL 0.1 M HCl solution is added to 100 mL 0.1M NaCl solution. Which of the following statements concerning the concentrations of species in the original NaCl solution is correct?

    (A) $[Na^+]$ increases
    (B) $[Na^+]$ does not change
    (C) $[OH^-]$ increases
    (D) $[Cl^-]$ increases
    (E) $[Cl^-]$ does not change

27. Which of the following reactions is NOT an oxidation–reduction reaction?

    (A) $H_2(g) + 1/2\ O_2(g) \rightarrow H_2O(l)$
    (B) $NaOH(aq) + HCl(aq) \rightarrow NaCl(aq) + H_2O(l)$
    (C) $H_2O_2(l) \rightarrow H_2O(l) + O_2(g)$
    (D) $CH_4(g) + O_2(g) \rightarrow H_2O(g) + CO_2(g)$
    (E) $Mg(s) + 2\ HCl(aq) \rightarrow MgCl_2(aq) + H_2(g)$

28. Which of the following aqueous solutions would have a pH greater than 7.0?

    (A) 0.5M HCl
    (B) 0.5M NH_4Cl
    (C) 0.5M HCN
    (D) 0.5M KCN
    (E) 0.5M H_2SO_4

29. $...LiAlH_4(s) + ...BCl_3(l) \rightarrow ...H_2H_6\ (g) + ...LiCl(s) + ...AlCl_3(s)$

    When the equation for the reaction represented above is balanced and all coefficients are reduced to lowest whole–number terms, the coefficient of B_2H_6 is:

    (A) 1
    (B) 2
    (C) 3
    (D) 4
    (E) 5

30. Which of the following equals to Avogadro's number EXCEPT?

    (A) the number of oxygen atoms in 3.01×10^{23} O_2 molecules
    (B) the number of helium atoms in 1 mole of helium gas
    (C) the number of Na^+ ions in 1 L of 1 M NaCl solution
    (D) the number of hydrogen atoms in 6.02×10^{23} molecules
    (E) the number of hydrogen atoms in 11.2 liters of hydrogen gas at STP

31. At constant temperature, the change of state of any substance from liquid to gas always includes which of the following?

    I. The breaking of covalent bonds
    II. An increase in the randomness of the system
    III. The absorption of energy

    (A) I only
    (B) II only
    (C) I and II only
    (D) II and III only
    (E) I, II, III

32. The heat absorbed when 10 g of a liquid is heated to rise the temperature from 25°C to 50°C is 2 kcals. What is the specific heat of the liquid (in cal/g°C)?

    (A) 4
    (B) 8
    (C) 12
    (D) 24
    (E) 48

GO ON TO THE NEXT PAGE

33. Raising the temperature at which a chemical reaction proceeds may do all of the following EXCEPT

    (A) Increase the molecule collision frequency
    (B) Increase the number of molecules with energy greater than the activation energy
    (C) Speed up the forward and reverse reactions
    (D) Decrease the randomness of the system
    (E) Increase average kinetic energy of reactants and procuds

34. What are the oxidation numbers of the atoms in $Fe_3(PO_4)_2$?

    (A) Fe: +2; P: +5; O: +2
    (B) Fe: +2; P: +7; O: +2
    (C) Fe: +3; P: +5; O: –2
    (D) Fe: +3; P: +7; O: –2
    (E) Fe: +2; P: +3; O: –2

35. Given a 6 liter sample of helium gas is at STP. If the gas is compressed to 4 liter at 15 degrees Celsius. What would be the pressure after the gas sample is compressed?

    (A) $760 \times \frac{6}{4} \times \frac{288}{273}$

    (B) $760 \times \frac{6}{4} \times \frac{273}{288}$

    (C) $760 \times \frac{4}{6} \times \frac{288}{273}$

    (D) $760 \times \frac{4}{6} \times \frac{273}{288}$

    (E) $760 \times \frac{6}{4} \times \frac{15}{0}$

36. Which of the following molecules is polar?

    (A) BH_3
    (B) NF_3
    (C) C_2H_6
    (D) SF_6
    (E) CCl_4

37. $C_2H_4(g) + 3O_2(g) \rightarrow 2CO_2(g) + 2H_2O(l)$

    When 100 grams of O_2 are allowed to react completely with 1.0 mole of C_2H_4 according to the equation above, which of the following occurs?

    (A) Some C_2H_4 remains unreacted.
    (B) Some O_2 remains unreacted.
    (C) More than 2 moles of H_2O is formed.
    (D) Less than 2 moles of CO_2 is formed.
    (E) The partial pressure of O_2 falls to zero.

38. A flask contains three times as many moles of H_2 gas as molecules of O_2 gas. If hydrogen and oxygen are the only gases present, what is the total pressure in the flask if the partial pressure due to hydrogen is P?

    (A) 4 P
    (B) 3 P
    (C) 4/3 P
    (D) 3/4 P
    (E) 5/4 P

39. 5 grams of crystalline NaOH is dissolved in 100 mL water. How many mL of a 0.5 M HCl solution are needed to completely neutralize the NaOH solution?

    (A) 25
    (B) 50
    (C) 100
    (D) 125
    (E) 250

40. Oxidation–reduction processes include all of the following EXCEPT

    (A) Cleaning up Hg spill with sulfur
    (B) Rusting of iron
    (C) Generating power by an alkaline battery
    (D) Combustion of heating oil
    (E) Using salt to melt snow

GO ON TO THE NEXT PAGE

41. Consider the gas phase equilibrium reaction:

    $H_2(g) + Br_2(g) \rightleftharpoons 2HBr(g)$

    with equilibrium constant, $K = 2.5 \times 10^3$, at 400°C.

    Is a system with 0.5 moles H_2, 0.4 moles Br_2 and 50 moles HBr at equilibrium?

    (A) Yes, the system is at equilibrium.
    (B) No, the reaction must shift to the left in to reach equilibrium.
    (C) No, the reaction must shift to the right in to reach equilibrium.
    (D) Br_2 will be completely consumed at equilibrium.
    (E) This system will never be at equilibrium.

42. Concerning the following equilibrium reactions

    I. $A + B \rightleftharpoons 2C + D$
    II. $2C + D \rightleftharpoons A + B$

    If the equilibrium constant for reaction I is 2.5×10^{-2}, what is the equilibrium constant for equilibrium reaction II?

    (A) 2.5×10^{-2}
    (B) 6.25×10^{-4}
    (C) $1/(2.5 \times 10^{-2})$
    (D) $1/(6.25 \times 10^{-4})$
    (E) 2.5×10^4

43. Which of the following approaches will convert 100 mL NaOH solution with pH = 13 to a solution of NaOH with pH = 12?

    (A) Adding distilled water to a total volume of 108 mL
    (B) Adding distilled water to a total volume of 500 mL
    (C) Adding distilled water to a total volume of 1 L
    (D) Adding 100 mL of 0.1 M HCl
    (E) Adding 100 mL of 0.1 M NaOH

44. Of the following ground state electron configurations, the one that represents the element of lowest first ionization energy (potential) is

    (A) $1s^2 2s^2 2p^5$
    (B) $1s^2 2s^2 2p^6$
    (C) $1s^2 2s^2 2p^6 3s^1$
    (D) $1s^2 2s^2 2p^6 3s^2$
    (E) $1s^2 2s^2 2p^6 3s^2 3p^1$

45. Potassium permanganate ($KMnO_4$) reacts with sodium sulfite (Na_2SO_3) in water to produce manganese dioxide (MnO_2), sodium sulfate (Na_2SO_4) and potassium hydroxide (KOH), as shown by the equation below:

    $2\ KMnO_4 + 3Na_2SO_3 \rightarrow$
    $\qquad\qquad 2MnO_2 + 3Na_2SO_4 + 2KOH$

    Which of the following is the reducing reagent in this reaction?

    (A) K^+
    (B) MnO_4^-
    (C) Na^+
    (D) SO_3^{2-}
    (E) H_2O

46. The first through sixth ionization energy (potential) of an element are 733, 1450, 7730, 10538, 13618, 18101 kJ/mole. An atom of this element is most likely to form an ion that has a charge of

    (A) -2
    (B) -1
    (C) $+1$
    (D) $+2$
    (E) $+3$

47. All of the following statement about ammonia, NH_3, are true EXCEPT

    (A) It has a characteristic odor.
    (B) It is a liquid at room temperature.
    (C) It is readily soluble in water.
    (D) Its aqueous solution has a pH great than 7.
    (E) It reacts readily with acid.

GO ON TO THE NEXT PAGE

48. What is the percent composition by mass of aluminum in aluminum sulfate, $Al_2(SO_4)_3$ (molar mass is 342)?

    (A) 12.5
    (B) 15.8
    (C) 22.4
    (D) 35.6
    (E) 43.1

49. Which of the following is responsible for the abnormally high boiling point of water?

    (A) Strong covalent bond in water molecules
    (B) Formation of hydrogen bonds between water molecules
    (C) High polarity of water molecules
    (D) High potential energy of water molecules
    (E) High molecular weight of water molecules

50. Some solid crystalline compounds slowly change to the gaseous state when left at room temperature in an open container. Which of the following is true about this phenomenon?

    (A) It is accompanied by increase in temperature.
    (B) It is accompanied by absorption of heat by the solid.
    (C) It is the result of a chemical reaction with air.
    (D) It is best described as fusion.
    (E) It is observed only with ice.

51. Concerning the reaction

    $C_3H_8 + O_2 \rightarrow CO_2 + H_2O$

    how many grams of O_2 will it take to completely combust 88 grams of C_3H_8?

    (A) 64
    (B) 172
    (C) 160
    (D) 224
    (E) 320

52. Which of the following combinations represents an element with a mass number of 75 and a net charge of +1?

    (A) 35 neutrons, 35 protons, 34 electrons
    (B) 40 neutrons, 40 protons, 39 electrons
    (C) 40 neutrons, 35 protons, 34 electrons
    (D) 37 neutrons, 38 protons, 39 electrons
    (E) 40 neutrons, 35 protons, 35 electrons

53. A physicist starts out with 32 grams of a radioactive element Z and after 28 hours he has only 2 grams left. What is the half–life of element Z?

    (A) 3.5 hours
    (B) 5.6 hours
    (C) 7.0 hours
    (D) 14.0 hours
    (E) 28.0 hours

54. The modern periodic table is arranged on the basis of

    (A) atomic mass
    (B) atomic radius
    (C) atomic charge
    (D) atomic number
    (E) number of neutrons

55. Which of the following oxides can dissolve in water to form a solution that would turn litmus indicator red in color?

    (A) Al_2O_3
    (B) K_2O
    (C) CO_2
    (D) CaO
    (E) ZnO

56. Which of the following pairs would make a buffer when equal number of mole dissolved in the same solution?

    (A) CH_3CO_2H and $Na^+CH_3CO_2^-$
    (B) HCl and NaOH
    (C) KOH and KCl
    (D) HBr and NaBr
    (E) H_2O and $NaNO_3$

GO ON TO THE NEXT PAGE

57. Which of the following has the highest electronegativity?

(A) Ca
(B) Cl
(C) Cs
(D) P
(E) Zn

58. Given the reaction

$$2Na(s) + Cl_2(g) \rightarrow 2NaCl(s) + 822 \text{ kilojoules}$$

How much heat is released if 0.6 mole of sodium reacts completely with chlorine?

(A) 61.7 kilojoules
(B) 123.3 kilojoules
(C) 246.6 kilojoules
(D) 453.2 kilojoues
(E) 906.4 kilojoules

59. What is indicated by the shape of the titration curve?

Volume of titrant added

A) A diprotic acid titrated with a strong base.
B) A triprotic acid titrated with a strong base.
C) A diprotic base titrated with a strong acid.
D) A triprotic base titrated with a strong acid.
E) A strong acid titrated with a strong base.

60. $HCl(aq) + Zn(s) \rightarrow ZnCl_2(aq) + H_2(g)$

In the reaction above, which term best describes the role of $HCl(aq)$?

(A) Brønsted acid
(B) Oxidizing reagent
(C) Reducing reagent
(D) Precipitate
(E) Cathode

61. How many moles of potassium ions are present in 2.50 L of 0.200 M potassium sulfate?

(A) 0.0800 mol
(B) 0.160 mol
(C) 0.400 mol
(D) 0.500 mol
(E) 1.00 mol

62. The reaction profile shown below is for an uncatalyzed reaction:

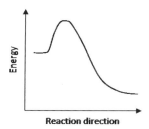

Which of the following is the reaction profile for the same reaction after the addition of a catalyst?

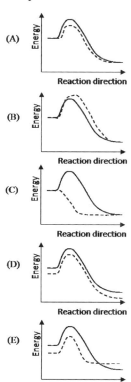

GO ON TO THE NEXT PAGE

$H_2(g) + I_2(g) + heat \rightleftharpoons 2HI(g)$

63. Which of the following changes to the equilibrium system above will increase the quantity of HI(g) in the equilibrium mixture?

    I. Adding $H_2(g)$
    II. Increase temperature
    III. Decrease the pressure

(A) I only
(B) III only
(C) I and II only
(D) II and III only
(E) I, II and III

64. When a fixed amount of gas has its Kelvin temperature and pressure doubled, what is the new volume of the gas relative to that of the original volume

(A) Four time of the original volume
(B) Twice of the original volume
(C) The same as the original volume
(D) One half of the original volume
(E) One fourth of the original volume

65. The electron configuration $1s^2 2s^2 2p^6 3s^2 3p^6$ represents the following EXCEPT

(A) Argon atom, Ar
(B) Chloride ion, Cl^-
(C) Potassium ion, Na^+
(D) Calcium ion, Ca^{2+}
(E) Aluminum ion, Al^{3+}

66. A chemist finds that there are 9 moles of oxygen in an aluminum (Al_2O_3, molar mass 102) sample. What is the total mass of this sample?

(A) 78.5 grams
(B) 144 grams
(C) 306 grams
(D) 918 grams
(E) 2754 grams

67. In which of the following compounds does bromine have the highest positive oxidation state?

(A) HBr
(B) BrF_3
(C) BrO_2
(D) NaBrO
(E) $NaBrO_3$

68. A solution is made by adding 4.0 grams of NaOH to enough water to make 1.0 liter of solution. What is the approximate pH of the resulting solution?

(A) 1
(B) 2
(C) 7
(D) 13
(E) 14

69. Neutralization of 500 mL of 0.5 M KOH requires the smallest volume of which of the following?

(A) 1M H_2SO_4
(B) 1 M CH_3COOH
(C) 3 M HCl
(D) 2 M NH_3
(E) 0.5 M H_3PO_4

70. Which of the following statements is the best expression for the sp^3 hybridization of carbon electrons?

(A) The new orbitals are one *s* orbital and three *p* orbitals
(B) The *s* electron is promoted to the *p* orbital
(C) The *s* orbital is deformed into a *p* orbital
(D) Four new and equivalent orbitals are formed
(E) The *s* orbital electron loses energy to fall back into a partially filled *p* orbital

STOP!

If you finish before time is called, you may check your work on this section only. Do not turn to any other section in the test.

Practice Test 6 Answers

#	Answer	#	Answer	#	Answer
1	E	25	C	49	B
2	C	26	E	50	B
3	A	27	B	51	E
4	B	28	D	52	C
5	D	29	B	53	C
6	C	30	D	54	D
7	E	31	D	55	C
8	A	32	B	56	A
9	D	33	D	57	B
10	B	34	C	58	C
11	D	35	A	59	C
12	A	36	B	60	B
13	C	37	B	61	E
14	D	38	C	62	A
15	B	39	E	63	C
16	C	40	E	64	C
17	A	41	B	65	E
18	D	42	C	66	C
19	B	43	C	67	E
20	C	44	C	68	D
21	E	45	D	69	C
22	D	46	D	70	D
23	A	47	B	71	
24	E	48	B	72	

Table title: PART A and C

#	Answer
101	True, True, No
102	True, True, No
103	False, False, No
104	False, True, No
105	True, False, No
106	True, True, Yes
107	False, True, No
108	True, False, No
109	False, True, No
110	False, True, No
111	True, True, Yes
112	True, True, Yes
113	False, True, No
114	True, True, Yes
115	True, False, No
116	

Table title: PART B

Calculation of the raw score

The number of correct answers: _____ = No. of correct

The number of wrong answers: _____ = No. of wrong

Raw score = No. of correct – No. of wrong x ¼ = _____

Score Conversion Table

Raw Score	Scaled Score	Raw Score	Scaled Score	Raw Score	Scaled Score
80	800	49	600	18	420
79	800	48	590	17	410
78	790	47	590	16	410
77	780	46	580	15	400
76	770	45	580	14	390
75	770	44	570	13	390
74	760	43	560	12	380
73	760	42	560	11	370
72	750	41	550	10	360
71	740	40	550	9	360
70	740	39	540	8	350
69	730	38	540	7	350
68	730	37	530	6	340
67	720	36	520	5	340
66	710	35	520	4	330
65	700	34	510	3	330
64	700	33	500	2	320
63	690	32	500	1	320
62	680	31	490	0	310
61	680	30	490	−1	310
60	670	29	480	−2	300
59	660	28	480	−3	300
58	660	27	470	−4	290
57	650	26	470	−5	280
56	640	25	460	−6	280
55	640	24	450	−7	270
54	630	23	450	−8	270
53	620	22	440	−9	260
52	620	21	440	−10	260
51	610	20	430		
50	600	19	420		

Explanations: Practice Test 6

1. **(E)** The atomic number of Ca and Cl are 20 and 17 respectively. After Ca loses 2 electrons and Cl gains 1 electron, they both have 18 electrons, and become Ca^{2+} and Cl^- respectively.

2. **(C)** Neutral Be atom has 4 electrons, and Be^{2+} has two electrons. Therefore, Be^{2+} has one more electron than neutral H atom.

3. **(A)** Neutral Al (Z = 13) and S (Z = 16) atoms have 13 and 16 electrons respectively.

4. **(B)** Neutral Li (Z = 3) and Na (Z = 11) has 3 and 11 electrons respectively, their +1 ions have 2 and 10 electrons respectively.

5. **(D)** N atom has 7 electrons, P^{3-} has 18 electrons (atomic number of P is 15).

6. **(C)** BeF_2 or other beryllium halogenate ($BeCl_2$, $BeBr_2$, BeI_2 etc.) are good examples of molecules with linear geometry. Be has only 2 valence electrons which are filled to two sp hybridized orbitals. The two sp hybridized orbitals are linear and form covalent bonds with two F atoms. There is no lone electron pair(s) with Be (there are three electron pairs with F). Therefore, the most stable configuration is linear.

7. **(E)** C atom in CH_4 has sp^3 hybridization, each sp^3 hybridized orbital form a sigma (σ) bond with one H atom. The geometry of CH_4 molecule is tetrahedral.

8. **(A)** N in NH_3 has sp^3 hybridization, however, since N has 5 valence electrons, one sp^3 hybridized orbital is filled with a pair of electrons, the rest sp^3 orbitals form sigma (σ) bond with H atom. The geometry of

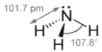

NH_3 molecule is trigonal pyramidal.

9. **(D)** Boron halogenates (BF_3, BCl_3, BBr_3, BI_3) are best examples of compounds with trigonal planar geometry. B has only 3 valence electrons filled to three sp^2 orbitals, which forms 3 sigma bonds with halogen atoms. There is no lone electron pair(s). Therefore, the most stable configuration is trigonal planar.

10. **(B)** B is called **triple point**, representing the temperature/pressure condition under which solid, liquid and gas can exist at the same time (**Figure** 1.1).

11. **(D)** D is called **critical point**, beyond which pressure alone cannot liquefy the gas.

12. **(A)** The process represented by A is change from solid to gas directly, and such process is called **sublimation**.

13. **(C)** Point C and all points on the line between B and D represent temperature/pressure condition under which liquid and vapor are at equilibrium and coexist.

14. **(D)** Line D represents vaporization, in which heat is absorbed when liquid water vaporizes at constant temperature (boiling point).

15. **(B)** Line D represents melting, in which heat is absorbed when solid melt at constant temperature (melting/freezing point).

16. **(C)** Line C represents heating of liquid from freezing (melting) point to boiling point.

17. **(A)** Line A represents heating of solid to melting point.

18. **(D)** $Ba(OH)_2$ solution is a strong base. Hydroxides of alkaline earth metals (Mg, Ca, Sr, Ba) are strong bases.

19. **(B)** H_3PO_4 solution is a strong acid.

20. **(C)** Acetic acid, CH_3COOH, an organic acid, is a weak acid.

21. **(E)** Ammonia solution is a weak base.

22. **(D)** Filtration is commonly used to separate solid from liquid.

23. **(A)** Distillation commonly used to separate mixture of two liquids based on the difference in their boiling points. When the mixture is heated in one container, the liquid with lower boiling point vaporizes first, and its vapor is condensed at lower temperature and collected into another container. The liquid with higher boiling point is left in the original container.

24. **(E)** In titration, an acid/base of known concentration is added to a base/acid with unknown concentration. When the acid/base of unknown concentration is completely neutralized, the volume of base/acid of known concentration are used to calculate the concentration of acid/base of unknown concentration.

25. **(C)** Fractional crystallization a method of separating substances in solution based on differences in solubility. For example, when the solution with mixed dissolved solids is heated, and the solvent is evaporated, the dissolved component with lower solubility will reach saturation first, and start to form crystal first.

26. **(E)** When 50 mL 0.1 M HCl is added into 100 mL 0.1 M NaCl, Na^+ is diluted to 0.1 M x $100/(100 + 50)$ = 0.067 M. Therefore, A and B are incorrect. NaCl solution is neutral with pH = 7, when a strong acid is added, it will be acidic, and $[H^+]$ increase, hence, $[OH^-]$ decreases. Choice C is incorrect. Since $[Cl^-]$ in both solutions are 0.1 M, its concentration will not change after mixing. Therefore, D is incorrect, and **E is correct answer**.

27. **(B)** B is an acid/base (neutralization) reaction, all other reactions are redox reactions.

28. **(D)** The salt KCN is formed between a strong base (KOH) and a weak acid (HCN), and its solution has a pH greater than 7 (basic). There are several guiding principles that help to determine the pH of a salt solution (Table 6.1):

Table 6.1 *pH of Salt*
a. Salts that are formed from strong bases and strong acids (e.g. NaCl, KNO_3 etc.) do not hydrolyze. The pH will remain neutral at 7. In general, salts containing halides (except F^- since HF is a weak acid) and an alkaline metals will dissociate into spectator ions.
b. Salts that are formed from strong bases and weak acids (e.g. KCN, NaF, K_2S) do hydrolyze giving pH greater than 7. The anion in the salt is derived from a weak acid (e.g. CN^-, F^-, CH_3COO^-, S^{2-}), and will accept the proton from the water in the reaction having the water act as an acid in this case leaving a hydroxide ion (OH^-). The cation will be from a strong base (e.g. alkaline or alkaline earth metals), it will dissociate into an ion and not affect the H^+. $CN^+ + H_2O \rightarrow HCN + OH^-$
c. Salts of weak bases and strong acids (e.g. NH_4Cl) do hydrolyze giving it a pH less than 7. This is due to the fact that the anion will become a spectator ion and fail to attract the H^+ ion, while the cation from the weak base will donate a proton to the water forming a hydronium ion. $NH_4^+ + H_2O \rightarrow NH_3 + H_3O^+$
d. Salts from a weak base and weak acid (e.g. NH_4F) also hydrolyze as the others, but more complex. The pH of such a solution will be determined by the strengths of both the weak acid and base which form the salt, and could be less than, equals to or greater than 7.

29. **(B)** The balanced equation is _3_ $LiAlH_4(s)$ + _4_ $BCl_3(l)$ → _**2**_ B_2H_6 (g) + _3_ $LiCl(s)$ + _3_ $AlCl_3(s)$.

30. **(D)** is correct answer. Hydrogen molecule is diatomic molecule, H_2. Therefore, 6.02×10^{23} hydrogen molecules contains $2 \times 6.02 \times 10^{23}$ hydrogen atoms, which is twice of Avogadro's number.

    (A) 3.01×10^{23} oxygen molecules have $2 \times 3.01 \times 10^{23} = 6.02 \times 10^{23}$ oxygen atoms.

    (B) Helium gas is monatomic (single atom) gas, and 1 mol of helium gas has 6.02×10^{23} atoms.

    (C) The concentration of Na^+ in 1 M NaCl solution is 1 M, and the number of mol of Na^+ in 1 L of 1 M Na^+ solution is 1M x 1L = 1 mol $Na^+ = 6.02 \times 10^{23}$ Na^+ ions.

    (E) At STP, 11.2 liters of hydrogen gas equals to (11.2 L)/(22.4 L/mol) = 0.5 mol H_2. Since hydrogen gas is diatomic molecule, 0.5 mol hydrogen gas has 1 mol (6.02×10^{23}) hydrogen atoms.

31. **(D)** When liquid is converted to gas, energy (heat) is needed to break the interaction between molecules (but not covalent bonds in a molecule, I is incorrect). That's why this process always needs energy (III is correct). Once liquid become gas, it occupies larger volume, the randomness also increases (II is correct).

32. **(B)** In the equation to calculate heat absorbed when heating a liquid, $q = mC_p\Delta T$, the unit of specific heat C_p depends on units of heat q and amount of liquid m. Rearrange the equation:

    $C_p = q/m\Delta T = 2$ kcal/(10 g x 25 °C) = 0.008 kcal/g °C = 8 cal/g °C.

33. **(D)** is correct answer. When increasing temperature of a reaction system, the average kinetic energy of all species (both reactants and products) in the system increases (statement E is correct); and there will be more molecules have energy higher than the activation energy (statement B is correct). The frequency of collision increases with speed of molecules (statement A is correct). The net effect is the rates of both forward and reverse reaction increase (statement C is correct). The system will becomes less organized (more chaotic), or randomness increases. Therefore, **statement D is incorrect**.

34. **(C)** For ionic compounds with oxyanion such as $Fe_3(PO_4)_2$, the charge of oxyanion equals to sum of O.N. of all atoms in the oxyanion. Charge of PO_4^{3-} is –3, and O.N. of O is –2, O.N. of P = $-3 - (-2 \times 4) = +5$.

    The sum of whole compound is 0; therefore, O.N. of Fe = $(0 - (-3 \times 2))/2 = +3$.

35. **(A)** Use ideal gas law, $P_1V_1 = nRT_1$ (1) and $P_2V_2 = nRT_2$ (2).

    Divide equation (1) by (2): $\frac{P_1V_1}{P_2V_2} = \frac{T_1}{T_2}$

    Rearrange the equation: $P_2 = P_1 \times \frac{V_1}{V_2} \times \frac{T_2}{T_1}$

    Plug in the numbers, $P_2 = 760 \times \frac{6}{4} \times \frac{288}{273}$. The correct answer is A.

    Note: temperature must be absolute temperature, and there is no need to covert volume to L.

36. **(B)** Both BF_3 and NH_3 have polar bond, their polarities are determined by the molecule geometry. BF_3 has trigonal planar geometry in which the polarities of the B–F bonds are cancelled, while NH_3 has trigonal pyramidal geometry (since N atom has one lone pair of electron), which is polar.

    (C) C_2H_6 is nonpolar because the molecule has symmetrical configuration, polarity of all C–H bonds is cancelled.

    (D) SF_6 is nonpolar since the molecule has octahedral geometry.

    (E) CCl_4 is nonpolar since the molecule has tetrahedral geometry like CH_4.

37. **(B)** For balanced reaction equation: $C_2H_4(g) + 3O_2(g) \rightarrow 2CO_2(g) + 2H_2O(l)$, follow steps below:

    Step 1: Calculate number of mol of O_2: (100 gram O_2)/(32 gram O_2/mol O_2) = 3.125 mol O_2.

    Step 2, Determine which reactant is limiting: since 1 mol C_2H_4 needs 3 mol of O_2 to react completely, there is extra O_2 (0.125 mol). **B is correct answer**. And partial pressure of O_2 will not become zero (E is incorrect).

    Step 3: Since C_2H_4 is limiting, there is exactly 2 mol CO_2 and H_2O produced. C and D are incorrect.

38. **(C)** Since H_2 is three times more than O_2, the partial pressure of O_2 is 1/3 of H_2, i.e. P/3. The total pressure is $P_{total} = P_{H2} + P_{O2} = P + P/3 = 4/3$ P.

39. **(E)** Follow the steps below:

   Step 1: The molar mass of NaOH is 40, the number of mol of NaOH = (5 gram NaOH)/(40 gram NaOH /mol NaOH) = 0.125 mol NaOH.
   Step 2: To completely neutralize 0.125 mol of NaOH, amount of H^+ needed is 0.125 mol (equivalent with 0.125 mol HCl).
   Step 3: Volume of 0.5 M HCl needed = (0.125 mol HCl)/(0.5 M) = 0.25 L = 250 mL.

   (**Note**: you don't need to calculate the concentration of NaOH solution. All you need is to calculate the number of mol of NaOH).

40. **(E)** Salt can help melting ice since when salt dissolved in water, it lower the melting point. There is no chemical reaction involved in this process. All other processes are chemical reactions involving transfer of electrons (reduction–oxidation reactions).

   (A) Cleaning up Hg spill with sulfur: $Hg(l) + S(s) \rightarrow HgS(s)$
   (B) Rusting of iron: $Fe + O_2 \rightarrow Fe_2O_3$
   (C) Generating power by an alkaline battery: $Zn(s) + 2MnO_2(s) \rightarrow ZnO(s) + Mn_2O_3(s)$
   (D) Combustion of heating oil: this is oxidation of hydrocarbon.

41. **(B)** $Q = [HBr]^2/[H_2][Br_2] = 50^2/(0.5 \times 0.4) = 1.25 \times 10^4 > K$. The reaction is not at equilibrium, and will shift to left.

42. **(C)** Reaction II is reverse of reaction I, take reciprocal of the equilibrium constant of reaction I, $1/(2.5 \times 10^{-2})$, which is equilibrium constant of reaction II.

43. **(C)** The concentration of OH^- in solution of pH = 13 and pH = 12 are 0.1 M and 0.01 M respectively. There are two way to reduce OH^- concentration from 0.1 M to 0.01 M. One way is to dilute the original solution by 10 times, i.e. dilute the solution from 100 mL to 1000 mL (1 L). Therefore, choice **(C) is correct answer**. The second way is adding acid to neutralize some OH^-. In choice (D), acid is added, but 100 mL 0.1 M HCl just neutralized all OH^- to bring the pH to 7. Choice E is incorrect since addition of the same solution (0.1 M NaOH, pH =13) will not change the concentration of H^+/OH^- and the pH.

44. **(C)** Element with the alkali metal type of electron configuration, i.e. there is one valence electron, has the lowest first ionization energy in a row. Choice C is such an electron configuration (Na).

45. **(D)** In this reaction, sulfite ion, SO_3^{2-}, convert to sulfate ion, SO_4^{2-}. Oxidation number of sulfur change from +4 to +6 by giving away 2 electrons. Therefore, SO_3^{2-} is reducing regent.

46. **(D)** There is a sharp increase from the second to third ionization energy, indicating this element has two valence electrons. Therefore, this element most likely loses two valence electrons to become a +2 ion.

47. **(B)** Ammonia is a gas, not liquid, at room temperature (boiling point –33°C).

48. **(B)** Percent mass of Al in $Al_2(SO_4)_3$ is ((2 x 27)/342) x 100% = 15.8%.

49. **(B)** Hydrogen bond between water molecules explains many abnormal properties of water, such as abnormal high melting and boiling points of water.

50. **(B)** This is a process called sublimation, which absorb heat for evaporation (B is correct). This is a phase change, temperature of solid will not change (A is incorrect). It is a physical process, no chemical reaction involved (C is incorrect). It's not fusion which is the movement of particles due to their random movement (D is incorrect), and can be observed with many substance (E is incorrect).

51. **(E)** See steps below:

Step 1: Balance the reaction equation: $C_3H_8 + 5O_2 \rightarrow 3CO_2 + 4H_2O$

Step 2: Calculate number of mol of **88 grams** of C_3H_8:

(88 gram C_3H_8)/(44 gram C_3H_8/mol C_3H_8) = 2 mol C_3H_8.

Step 3: Calculate number of mol of O_2 needed: 2 mol C_3H_8 x (5 mol O_2/1 mol C_3H_8) = 10 mol O_2.

Step 4: Calculate mass of O_2 needed: (10 mol O_2) x (32 gram O_2/mol O_2) = 320 gram O_2.

Combine calculation: $88 \; \text{grams of C}_3\text{H}_8 \times \dfrac{1 \; \text{mol C}_3\text{H}_8}{44 \; \text{gram C}_3\text{H}_8} \times \dfrac{5 \; \text{mol O}_2}{1 \; \text{mol C}_3\text{H}_8} \times \dfrac{32 \; \text{gram O}_2}{1 \; \text{mol O}_2} = 320$ gram O_2.

52. **(C)** Choice C has sum of neutrons and protons of 75 (mass number) and one less electron than proton (+1 of charge).

53. **(C)** Mass of the radioactive isotope change from 32 grams to 2 grams, or reduced by 16 times; four half–lives must have passed ($16 = 2^4$). Therefore, the half–life is 28 hour/4 = 7 hours.

54. **(D)** The elements in modern periodic table are arranged by the increasing number of their atomic numbers (number of protons). Atomic number determines the electron configuration of an element, which further determines the chemical and physical properties of the element. Elements in periodic table are arranged to rows and columns. In each row (period), the elements has same number of electron shells, while in each column (group), element has similar electron configurations of the outmost layer.

55. **(C)** Litmus indicator paper is red in acids and blue in bases. Al_2O_3, K_2O, CaO, ZnO are basic oxides while CO_2 is the acidic oxide. When CO_2 dissolves in water, it reacts with water to form carbonic acid. Only carbon dioxide turns blue litmus red (see **Table 5.1**).

56. **(A)** is the only option which contains a weak acid and its conjugate base, and such solution makes good buffer solution. All other pairs consist of strong acid, strong base or salt which are result of neutralization reactions between strong acid and strong base.

57. **(B)** This question is about periodic property (electronegativity) of elements. Generally, in the same row (period), electronegativity increases from left to right; therefore, Cl has higher electronegativity than P. Second, within a group, electronegativity decrease downward. Use these two rules together, Cl has higher electronegativity than Zn, Ca, and Cs (**Figure** 6.1).

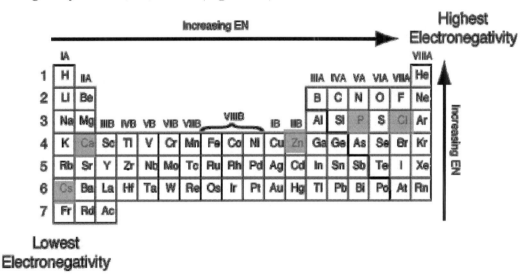

Figure 6.1 Periodic trends of electronegativity.

58. **(C)** Calculation of stoichiometry with enthalpy is similar with regular stoichiometry calculation

Heat released = (822 kJ/2 mol Na) x (0.6 mol Na) = 246.6 kJ.

59. (**C**) The titration curve is from a diprontic base which is titrated by a strong acid (**Table** 6.2).

For acid/base titration, a titration curve usually provides three types of information.

a. Acid/base of original solution and titrant. If pH increase with volume of titrant, this is an acid titrated by a base. Otherwise, it's a base titrated by acid.

b. How many H^+ or OH^- in formula of original acid or base. The sharp increase or decrease in pH on an acid/base titration curve indicates titration endpoint. If there is one endpoint, the original acid/base has only one H^+ or OH^- (monoprotic). If there are two endpoints, the original acid/base has two H^+ or OH^- available (diprotic).

c. Shape of curve provides information on strength of acid/base of solution titrated (not further discussed for SAT Chemistry subject test).

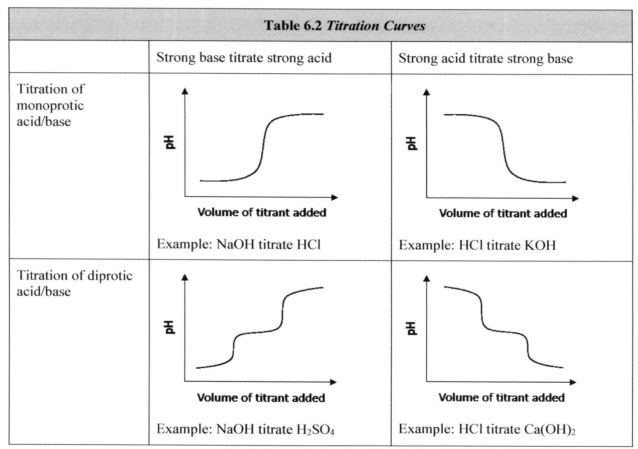

Table 6.2 *Titration Curves*		
	Strong base titrate strong acid	Strong acid titrate strong base
Titration of monoprotic acid/base	Example: NaOH titrate HCl	Example: HCl titrate KOH
Titration of diprotic acid/base	Example: NaOH titrate H_2SO_4	Example: HCl titrate $Ca(OH)_2$

60. (**B**) In the reaction $HCl(aq) + Zn(s) \rightarrow ZnCl_2(aq) + H_2(g)$, hydrogen atom in HCl gain electron to be reduced to H_2, and therefore, HCl is an oxidizing reagent.

61. (**E**) The number of mol of K_2SO_4 in 2.50 L of 0.20 M solution is (2.50 L) x (0.20 M) = 0.50 mol K_2SO_4.

The number of mol of K^+ in this solution is 0.50 mol x 2 = 1 .00 mol K^+.

62. (**A**) is correct answer. Catalyst change activation energy of both reactant and product, but not potential energy of reactant and product.

63. (**C**) This problem is about **Le Chatelier's Principle** (**Table** 1.5), which states that if a chemical system at equilibrium experiences a change in concentration, temperature, volume, or partial pressure, then the equilibrium shifts to counteract the imposed change and a new equilibrium is established.

I. Based on this principle, adding H_2 will cause equilibrium shift to the right, and increase HI produced.

II. Since the reaction absorb heat, increase in temperature will also cause the equilibrium shift to right, and increase production of HI.

III. Pressure does not has impact on the equilibrium, since total number of mole does not change in the equilibrium reaction.

64. **(C)** Volume of a gas is in proportion with Kelvin temperature, and in reverse proportion with pressure, as shown by ideal gas law, $V = nRT/P$. When both Kelvin temperature and pressure are doubled, the volume will remain unchanged since the impacts of the changes in T and P are cancelled.

65. **(E)** The atomic number of Al is 13, after losing 3 valence electrons and become Al^{3+} which has 10 electrons left. The electron configuration of Al^{3+} is $1s^2 2s^2 2p^6$ (same as neon).

66. **(C)** Follow the steps below:

Step 1: Number of mol of Al_2O_3 = 9 mol O x (1 mol Al_2O_3/3 mol O) = 3 mol Al_2O_3
Step 2: Mass of Al_2O_3 = 3 mol Al_2O_3 x (102 g Al_2O_3/ mol Al_2O_3) = 306 g Al_2O_3

67. **(E)** The oxidation number of Br in $NaBrO_3$ is +5. See **Table** 1.11 for assigning oxidation number. Oxidation numbers in other compounds are as follow:

(A) HBr –1
(B) BrF_3 +3
(C) BrO_2 +2
(D) NaBrO +1

68. **(D)** NaOH is a strong base. When dissolved in water, NaOH completely dissociate and OH^- is released to solution.

Step 1: Number of mol of NaOH: 4.0 gram/(40 gram/mol) = 0.1 mol NaOH
Step 2: Concentration of NaOH solution: 0.1 mol NaOH/1 L = 0.1 M.
Step 3: Concentration of OH^- = 0.1 M. Concentration of H^+ = $10^{-14}/0.1 = 10^{-13}$ M.
Step 4: pH of the solution = $-\log(10^{-13})$ = 13.

69. **(C)** is correct answer.

Number of mol of OH^- in KOH solution = 0.5 L x 0.5 M = 0.25 mol OH^-
(A) Volume needed of 1 M H_2SO_4 = 0.25 mol / (2 x 1 M) = 0.125 L
(B) Volume needed of 1 M CH_3COOH = 0.25 mol / (1 x 1 M) = 0.25 L
(C) Volume needed of 3 M HCl = 0.25 mol / (1 x 3 M) = 0.083 L
(D) NH_3 is a base, cannot be used to neutralize base.
(E) Volume needed of 0.5 M H_3PO_4 = 0.25 mol / (3 x 0.5 M) = 0.17 L

70. **(D)** Hybridized orbitals (e.g sp^3) are different from orbitals which are used to hybridize (e.g. s and p). They are equivalent. The energy level of hybridized orbitals is in between the orbitals used to hybridize.

101. **Correct answer: I True, II True, CE No**

Explanation: 1.0 mol of H_2SO_4 contains twice as much of H^+ ion as OH^- ion in 1.0 mol NaOH; therefore, NaOH is limiting reagent (Statement I is correct). Molar mass of H_2SO_4 is 98, and molar mass of NaOH is 40; therefore, Statement II is correct. However, Statement II cannot explain Statement I, because molar mass alone does not determine which reactant is limiting; numbers of mole of the reactants (H^+ and OH^-) in this neutralization reaction determine which reactant is limiting.

102. **Correct answer: I True, II True, CE No**

Explanation: The volume of a gas decreases when temperature decreases, and/or when pressure increases. Therefore, at low temperature and high pressure, the volume of the gas is reduced, and tends to condense (Statement I is correct). The interaction between molecules in a gas is very low, the molecules tend to move freely, and fill the container easily (Statement II is correct). However, Statement II does not explain Statement I, because the tendency of a gas to fill the container has nothing with how the gas is changed when temperature and pressure change.

103. **Correct answer: I False, II False, CE No**

Explanation: The empirical formula of H_2O and H_2O_2 are H_2O and HO respectively (Statement I is false). The oxidation numbers of O atom in H_2O and H_2O_2 are -2 and -1 respectively, the oxidation numbers of H atom are the same ($+1$); therefore, Statement II is false.

104. **Correct answer: I False, II True, CE No**

Explanation: In a group, the tendency to gain electron of the elements decreases downward; or the oxidizing strength decreases. Therefore, chlorine is stronger oxidizing reagent than bromine (Statement I is false). Second, the atomic radius increases down a column (group), and bromine atom is larger than chlorine (Statement II is true).

105. **Correct answer: I True, II False, CE No**

Explanation: Combustion of hydrocarbon releases large amount of heat, it's exothermic reaction (Statement I is true). Exothermic reaction releases heat, no heat is needed once the exothermic reaction is started (Statement II is false).

106. **Correct answer: I True, II True, CE Yes**

Explanation: Silicon atom has 4 valence electrons, and forms for identical Si–Cl bonds. The $SiCl_4$ molecule has symmetric tetrahedral geometry, which cancel all polarity of the polar Si–Cl bonds. Therefore, $SiCl_4$ molecule is nonpolar. Statement I and II are true, and Statement II correctly explains Statement I.

107. **Correct answer: I False, II True, CE No**

Explanation: Temperature remains constant when the state of a substance changes (from solid to liquid, or from liquid to gas etc.) even though heat are continuously added or released (Statement I is false). Temperature is a measure of kinetics energy of particles in a system, average kinetic energy increases with temperature (statement II is true).

108. **Correct answer: I True, II False, CE No**

Explanation: Magnesium is alkali earth metal (Group II) with 2 valence electrons, it will lose these valence electrons when react with nonmetal elements such as chlorine or oxygen. For chlorine, it has 7 valence electrons and tends to gain one electron to satisfy the octet rule (Statement II is false). Therefore, the formula of the compound formed between Mg and Cl is $MgCl_2$ (Statement I is true).

109. **Correct answer: I False, II True, CE No**

Explanation: Both the radius and atomic number increase down a column (group, statement I is false and statement II is true).

110. **Correct answer: I False, II True, CE No**

Explanation: When temperature of a gas decreases, its volume also decreases (Charles' law). Statement I is false. When pressure of a gas decreases, its volume increases (Boyle's law). Statement II is true. There is no cause–effect relationship between the two statements.

111. **Correct answer: I True, II True, CE Yes**

Explanation: The lighter the gas molecule, the faster its effusion rate (effusion rate is in reverse proportion to the square root of molar mass). Therefore diatomic nitrogen effuses faster than diatomic oxygen, and Statement II correctly explains Statement I.

112. **Correct answer: I True, II True, CE Yes**

   Explanation: In reduction reaction, the reactant (oxidizing reagent) gains electron(s); in contrast, in oxidation reaction, the reactant (reducing reagent) loses electron(s). Therefore, both Statements I and II are true, and Statement II correctly explains Statement I.

113. **Correct answer: I False, II True, CE No**

   Explanation: Water can act as both an acid and a base, and it's amphoteric (Statement II is true). But water is not good buffer since it does not resist pH change (Statement I is false).

114. **Correct answer: I True, II True, CE Yes**

   Explanation: "Like dissolves like" refers to polarity of solvent and solutes. Polar solvents dissolve polar solutes, and nonpolar solvents dissolve nonpolar solutes. Carbon tetrachloride molecules are nonpolar, and water molecules are polar. Therefore, carbon tetrachloride is insoluble in water (Statement I is true). The reason why water molecules are polar is because their geometry, asymmetric configuration (Statement II is true). Statement II explains why water molecules are polar, and why water and carbon tetrachloride are not insoluble in each other.

115. **Correct answer: I True, II False, CE No**

   Explanation: When a solid dissolved in solvent, the randomness increases (the system is less organized), i.e. the entropy of the system increased (Statement I is true). However, heat is not always released in dissolution (Statement II is false). For some salts, heat is absorbed when they are dissolved in water. This is because the dissolution is a 2–step process, one step is to dissociate the bonds which hold the solid together, this process absorbs energy (endothermic); the second step is formation of the bonds between solutes (molecule or ion) and solvent, this process releases heat (exothermic). The overall impact of dissolution depends on which one is larger.

Made in the USA
Middletown, DE
20 December 2018